Physiology by Numbers

An Encouragement to Quantitative Thinking

SECOND EDITION

RICHARD F. BURTON
University of Glasgow, Glasgow

PUBLISHED BY THE PRESS SYNDICATE OF THE UNIVERSITY OF CAMBRIDGE
The Pitt Building, Trumpington Street, Cambridge, United Kingdom

CAMBRIDGE UNIVERSITY PRESS
The Edinburgh Building, Cambridge, CB2 2RU, UK   http://www.cup.cam.ac.uk
40 West 20th Street, New York, NY 10011–4211, USA   http://www.cup.org
10 Stamford Road, Oakleigh, Melbourne 3166, Australia
Ruiz de Alarcón 13, 28014 Madrid, Spain

First published 1994
Second edition published 2000

Printed in the United Kingdom at the University Press, Cambridge

Typeface Utopia 9.25/13.5 and Meta Plus.   *System* QuarkXPress® [SE]

A catalogue record for this book is available from the British Library

Library of Congress Cataloguing in Publication Data

Burton, R. F. (Richard F.)
  Physiology by numbers: an encouragement to quantitative thinking
  / Richard F. Burton.
     p.   cm.
  Includes bibliographical references and index.
  ISBN 0 521 77200 1 (hb.). – ISBN 0 521 77703 8 (pbk.)
  1. Human physiology – Mathematics Problems, exercises, etc.
I. Title.
QP40.B98   2000
612′.001′51–dc21   99-16237 CIP

ISBN 0 521 77200 1 hardback
ISBN 0 521 77703 8 paperback

CONTENTS

PREFACE TO THE SECOND EDITION

When I started to write the first edition of this book, I particularly had in mind readers somewhat like myself, not necessarily skilled in mathematics, but interested in a quantitative approach and appreciative of simple calculations that throw light on physiology. In the end I also wrote, as I explain more fully in my original Preface, for those many students who are ill at ease with applied arithmetic. I confess now that, until I had the subsequent experience of teaching a course in 'quantitative physiology', I was not fully aware of the huge problems so many present-day students have with this, for so many are reluctant to reveal them. Part of my response to this revelation was *Biology by Numbers* (Burton 1998), a book which develops various simple ideas in quantitative thinking while illustrating them with biological examples. In revising *Physiology by Numbers*, I have retained the systematic approach of the first edition, but have tried to make it more accessible to the number-shy student. This has entailed, amongst other things, considerable expansion of the first chapter and the writing of a new chapter to follow it. In particular, I have emphasized the value of including units at all stages of a calculation, both to aid reasoning and to avoid mistakes. I should like to think that the only prior mathematics required by the reader is simple arithmetic, plus enough algebra to understand and manipulate simple equations. Logarithms and exponents appear occasionally, but guidance on these is given in Appendix B. Again I thank Dr J. D. Morrison for commenting on parts of the manuscript.

R. F. Burton

Let us therefore take it that in a man the amount of blood pushed forward in the individual heartbeats is half an ounce, or three drams, or one dram, this being hindered by valves from re-entering the heart. In half an hour the heart makes more than a thousand beats, indeed in some people and on occasion, two, three or four thousand. Now multiply the drams and you will see that in one half hour a thousand times three drams or two drams, or five hundred ounces, or else some such similar quantity of blood, is transfused through the heart into the arteries – always a greater quantity than is to be found in the whole of the body.

But indeed, if even the smallest amounts of blood pass through the lungs and heart, far more is distributed to the arteries and whole body than can possibly be supplied by the ingestion of food, or generally, unless it returns around a circuit.

William Harvey, De Motu Cordis, *1628 (from the Latin)*

In more familiar terms, if the heart beats, say, 70 times a minute, ejecting 70 ml of blood into the aorta each time, then more fluid is put out in half an hour (147 l) than is either ingested in that time or contained in the whole of the body. Therefore the blood must circulate. Thus may the simplest calculation bring understanding. I invite the reader to join me in putting two and two together likewise, hoping that my collection of simple calculations will also bring enlightenment.

Although my main aim is to share some insights into physiology obtained through calculation, I have written also for those many students who seem to rest just on the wrong side of an educational threshold – knowing calculators and calculus, but shy of arithmetic; drilled in accuracy and unable to approximate; unsure what to make of all those physiological concentrations, volumes and pressures that are as meaningless as telephone numbers until toyed with, combined, or re-expressed. As 'an encouragement to quantitative thinking' I also offer, for those ill at ease with arithmetic, guidance on how to cheat at it, cut corners, and not be too concerned for spurious accuracy. Harvey's calculations illustrate very well that a correct conclusion may be reached in spite of considerable inaccuracy. In his case it was the estimate of cardiac output that was wrong; it is now known to be about two and a half ounces per beat. (There are eight drams to the ounce.)

Much of physiology requires precise computation, so I must not appear too much the champion of error and slapdash. There are, however, situations where even the roughest of calculations may suffice. Consider the generalization (see Section 3.10) that small mammals have higher metabolic rates per unit body mass than do large ones: taking the case of a hypothetical mouse with the relative metabolic rate of a steer, Max Kleiber (1961) calculated that to keep in heat balance in an environment at 3 °C its surface covering, if like that of the steer, would need to be at least 20 cm thick! Arguments of this kind appear below. Be warned, however, that improbable answers are not always wrong, as exemplified by Rudolph Heidenhain's calculation of glomerular filtration rate in 1883 (Section 6.5).

The book is based on an assortment of questions to be answered by calculation, together with some introductory and background information and comment on the answers. (The answers are given at the back of the book, together with notes and references.) Such a quantitative approach is more suited to some areas of physiology than to others and the coverage of the book naturally reflects this. The book is neither a general guide to basic physiology, nor a collection of brain-teasers or practice calculations. It rarely strays from shopkeeper's arithmetic and it is not a primer of mathematical physiology or of mathematics for physiologists. Rather, it is supplementary thinking for those who have done, or are still doing, at least an elementary course in Physiology. I have learned much myself from the calculations and hope that other mature students may learn from them too.

Except where otherwise stated, the calculations refer to the human body. This is often taken as that of the physiologist's standard 70-kg adult man and many 'standard', textbook quantities are used here. This is partly to reinforce them in the reader's memory and build bridges from one to another, but such standard values are also a natural starting point for back-of-envelope calculations. Indeed, if there is any virtue to learning these quantities, it is surely helpful to exercise them and put them to use. Thus may one hope to bring life to numbers – and not just numbers to Life.

The link between the learning and usefulness of quantities may be viewed the other way round. A student may memorize many of them for examinations and for future clinical application, but which are most profitably learnt for the better understanding of the body? Those with most uses? In how many elementary contexts is it helpful to know the concentration of sodium in extracellular fluid? Is that of magnesium as useful? Or manganese? Such questions of priority are as important for those inclined to overtax their memories unreasonably as for the lazy. This book may help both with these decisions and with the learning process itself.

Partly for reasons just indicated, many of my 'numbers' come from textbooks. Working on this text, however, I came increasingly to realize how hard it may be to find what one supposes to be well-known quantities. Textbooks have less and less room for these as other knowledge accumulates, of course, and there is a laudable tendency for concepts to displace quantitative detail. So do not disdain the older books! Diem (1962) has been a very useful source. Sometimes when a quantitative argument seems frustrated through lack of reliable figures, the solution is to turn it on its head, depart from the natural sequence of calculations, and defer the uncertainties to the end. The reader may spot where I was able to rescue items that way. Only once have I resorted to original data; I am very grateful to Dr Andrew Chappell for dissecting and weighing human muscles for me (Section 9.4).

I thank also all my colleagues who read portions of draft manuscript or otherwise gave of their time and wisdom, and in particular Dr F. L. Burton, Professor J. V. G. A. Durnin, Dr M. Holmes, Dr O. Holmes, Professor S. Jennett, Dr D. J. Miller, Dr J. D. Morrison, Dr G. L. Smith and Dr N. C. Spurway.

R. F. Burton

Understand the objectives as stated in the Preface to the first edition; be clear what the book is – and what it is not. Since it is written for readers of widely varying physiological knowledge and numerical skills, read selectively. Chapters 3–9, and their individual subsections, need not be read in sequence.

Although the book is primarily about physiology, another objective is to encourage and facilitate quantitative thinking in that area. If such thinking does not come easily to you, pay particular attention to Chapter 1. Note too that the calculations are not intended to be challenging. Indeed, many are designed for easy mental, or back-of-envelope, arithmetic – and help is always to hand at the back of the book, in 'Notes and Answers'. The notes often deal with points considered either too elementary or too specialized for the main text.

Consider carefully the validity of all assumptions and simplifications. If you try guessing answers before calculating them, you are more likely to be rewarded, in some cases, with a surprise.

If you are unfamiliar with exponents or logarithms, note the guidance given in Appendix B. The mathematics of exponential time courses are not dealt with in a single place, but most of the essentials are covered incidentally (see pages 13–16, 80–81, 98–100, 210–211, 219).

1 Introduction to physiological calculation: approximation and units

One purpose of the many calculations in later chapters is to demonstrate, as 'an encouragement to quantitative thinking', that a little simple arithmetic can sometimes give useful insights into physiology. Encouragement in this chapter takes the form of suggestions for minimizing some of the common impediments to calculation. I have mainly in mind the kinds of arithmetical problem that can suggest themselves outside the contexts of pre-planned teaching or data analysis. Some of the ideas are elementary, but they are not all as well known as they should be. Much of the arithmetic in this book has deliberately been made easy enough to do in the head (and the calculations and answers are given at the back of the book anyway). However, it is useful to be able to cut corners in arithmetic when a calculator is not to hand and guidance is first given on how and when to do this. Much of this chapter is about physical units, for these have to be understood, and casual calculation is too easily frustrated when conversion factors are not immediately to hand. It is also true that proper attention to units may sometimes propel one's arithmetical thinking to its correct conclusion. Furthermore, analysis in terms of units can also help in the process of understanding the formulae and equations of physiology, and the need to illustrate this provides a pretext for introducing some of these. The chapter ends with a discussion of ways in which exponents and logarithms come into physiology, but even here there is some attention to the topics of units and of approximate calculation.

1.1 Arithmetic – speed, approximation and error

We are all well drilled in accurate calculation and there is no need to discuss that; what some people are resistant to is the notion that accuracy may sometimes take second place to speed or convenience. High accuracy in physiology is often unattainable anyway, through the inadequacies of data. These points do merit some discussion. Too much initial concern for accuracy and rigour should not be a deterrent to calculation, and those people who confuse the precision of their calculators with accuracy are urged to cultivate the skills of approximate ('back-of-envelope') arithmetic. Discussed here are these skills, the tolerances implicit in physiological variability, and at times the necessity of making simplifying assumptions.

On the matter of approximation, one example should suffice. Consider the following calculation:

$$311/330 \times 480 \times 6.3.$$

A rough answer is readily obtained as follows:

$$\text{(nearly 1)} \times \text{(just under 500)} \times \text{(just over 6)}$$
$$= \text{slightly under 3000.}$$

The 480 has been rounded up and 6.3 rounded down in a way that should roughly cancel out the resulting errors. As it happens, the error in the whole calculation is only 5%.

When is such imprecision acceptable? Here is something more concrete to be calculated: *In a man of 70 kg a typical mass of muscle is 30 kg: what is that as a percentage?* An answer of 42.86% is arithmetically correct, but absurdly precise, for the mass of muscle is only 'typical', and it cannot easily be measured to that accuracy even with careful dissection. An answer of 43%, even 40%, would seem precise enough.

Note, in this example, that the two masses are given as round numbers, each one being subject both to variation from person to person and to error in measurement. This implies some freedom for one or other of the masses to be changed slightly and it so happens that a choice of 28 kg, instead of 30 kg, for the mass of muscle would make the calculation easier. Many of the calculations in this book have been eased for the reader in just this way.

Rough answers will often do, but major error will not. Often the easiest mistake to make is in the order of magnitude, i.e. the number of noughts or the position of the decimal point. Here again the above method of approximation is useful – as a check on order of magnitude when more accurate arithmetic is also required. Other ways of avoiding major error are discussed in Section 1.3.

Obviously, wrong answers can be obtained if the basis of a calculation is at fault. However, some degree of simplification is often sensible as a first step in the exploration of a problem. Many of the calculations in this book involve simplifying assumptions and the reader would be wise to reflect on their appropriateness; there is sometimes a thin line between what is inaccurate, but helpful in the privacy of one's thoughts, and what is respectable in print. Gross simplification can indeed be helpful. Thus, the notion that the area of body surface available for heat loss is proportionately less in large than in small mammals is sometimes first approached, not without some validity, in terms of spherical, limbless bodies. The word 'model' can be useful in such contexts – as a respectable way of acknowledging or emphasizing departures from reality.

1.2 Units

Too often the simplest physiological calculations are hampered by the fact that the various quantities involved are expressed in different systems of units for which interconversion factors are not to hand. One source of information may give pressures in mmHg, and another in cmH_2O, Pa ($= N/m^2$) or $dyne/cm^2$ – and it may be that two or three such diverse figures need to be combined in the calculation. Spontaneity and enthusiasm suffer, and errors are more likely.

One might therefore advocate a uniform system both for physiology generally and for this book in particular – most obviously the metric *Système International d'Unité* or SI, with its coherent use of kilograms, metres and seconds. However, even if SI units are universally adopted, the older books and journals with non-SI units will remain as sources of quantitative information (and one medical journal, having tried the exclusive use of SI units, abandoned it). This book favours the units that seem most usual in current textbooks and in hospitals and, in any case, the reader is not required to struggle with conversion factors. Only occasionally is elegance lost, as when, in Section 5.10, the law of Laplace, so neat in SI units, is re-expressed in other terms.

Table 1.1 lists some useful conversion factors, even though they are not much needed for the calculations in the book. Rather, the table is for general reference and 'an encouragement to (other) quantitative thinking'. For the same reason, Appendix A supplies some additional physical, chemical and mathematical quantities that can be useful to physiologists. Few of us would wish to learn all of Table 1.1, but, for reasons explained below, readers with little physics should remember that $1 N = 1 kg m/s^2$, that $1 J = 1 N m$ and that $1 W = 1 J/s$. The factor for converting between calories and joules may also be worth remembering, although '4.1855' could be regarded as over-precise for

Table 1.1. *Conversion factors for units*

Time		
1 day (d)	**86,400 s**	1440 min
Distance		
1 metre (m)	39.4 inch	
1 foot	0.305 m	
1 km	0.621 mile	
1 Ångstrom unit	0.1 nanometre (nm)	
Volume		
1 litre (l)	**10^{-3} m^3**	1 dm^3
Velocity		
1 mph	**0.447 m/s**	1.609 km/h
Acceleration (gravitational)		
g	**9.807 m/s^2**	32.17 ft/s^2
Mass		
1 lb	**0.4536 kg**	16 oz (avoirdupois)
Force		
1 newton (N)	**1 kg m/s^2**	102 g-force
1 kg-force	9.807 N	1 kilopond
1 dyne	10^{-5} N	1 g cm/s^2
Energy		
1 joule (J)	**1 N m**	
1 erg	10^{-7} N m	1 dyne cm
1 calorie (cal)	4.1855 J	
1 m kg-force (1 kg m)	9.807 J	
Power		
1 watt (W)	**1 J/s**	860 cal/h
Pressure and stress		
1 N/m^2	1 pascal (Pa)	
1 kg-force/m^2	9.807 N/m^2	1 mmH$_2$O
1 torr	1 mmHg	13.6 mmH$_2$O
1 mmHg	133.3 N/m^2	0.1333 kPa
750 mmHg	100.0 kN/m^2	
1 atmosphere	101.3 kN/m^2	760 mmHg

Note: SI units, fundamental or derived, are in **bold** lettering.

most purposes. In a similar vein, the '9.807' can often be rounded to '10', but
it is best written to at least two significant figures (9.8) since, especially
without units, its identity is then more apparent than that of commonplace
'10'. It helps to have a feeling for the force of 1 N in terms of weight; it is
approximately that of a 100-g object – Newton's legendary apple perhaps. As
for pressure, 1 kg-force/m² and 9.807 N/m² may be better appreciated as
1 mmH$_2$O, which is perhaps more obviously small.

   Units may be written, for example, in the form m/s² or m s^{-2}. I have chosen
what I believe to be the more familiar style. The solidus (/) may be read as
'divided by' or as 'per', and often these meanings are equivalent. However,
there is the possibility of ambiguity when more than one solidus is used, and
that practice is best avoided. We shortly meet (for solubility coefficients) a
combination of units that can be written unambiguously as 'mmol/l per
mmHg', 'mmol/l mmHg', 'mmol/(l mmHg)' and 'mmol l^{-1} mmHg^{-1}'. What is
ambiguous is 'mmol/l/mmHg', for if each solidus is read as 'divided by'
rather than as 'per', then the whole combination would be wrongly read as
'mmol mmHg/l'. In the course of calculations, e.g. involving the cancellation
of units (see below), it can be helpful to make use of a horizontal line to indicate division, so that 'mmol/l per mmHg' becomes:

$$\frac{\text{mmol/l}}{\text{mmHg}} \quad \text{or} \quad \frac{\text{mmol}}{\text{l mmHg}}.$$

1.3 How attention to units can ease calculations, prevent mistakes and provide a check on formulae

Students often quote quantities without specifying units, thereby usually
making the figures meaningless. All know that units and their interconver-
sions have to be correct, but the benefits of keeping track of units when calculating are not always fully appreciated. Thus, their inclusion in all stages of
a calculation can prevent mistakes of various kinds. Indeed, attention to
units can sometimes lead to correct answers (e.g. when tiredness makes
other reasoning falter), or help in checking the correctness of half-remembered formulae. Too many people flounder for lack of these simple notions.
The illustrations that follow involve commonplace physiological formulae,
but if some of them are unfamiliar that could even help here, by making the
usefulness of the approach more apparent. The formulae are in a sense incidental, but, since they are useful in their own right, the associated topics are
highlighted in bold type.

To illustrate the approach I start with an example so simple that the benefits of including units in the calculation may not be apparent. It concerns the excretion of urea. An individual is producing urine at an average rate of, say, 65 ml/h. The average concentration of urea in the urine is 0.23 mmol/ml. The rate of urea excretion may be calculated as the product of these quantities, namely 65 ml/h × 0.23 mmol/ml. The individual units (ml, mmol and min) are to be treated as algebraic quantities that can be multiplied, divided or cancelled as appropriate. Therefore, for clarity, the calculation may be written out thus:

$$65\,\frac{ml}{h} \times 0.23\,\frac{mmol}{ml} = 15\,\frac{mmol}{h}, \text{ i.e. 15 mmol/h.}$$

With the units spelt out like that, it would immediately become apparent if, say, there were an inappropriate mixing of volume units, e.g. millilitres in 'ml/h' with litres in 'mmol/l'. (What would then need to be done is probably obvious, but there is one particular kind of procedure for introducing conversion factors – in this case the '1000' relating ml to l – that can be helpful when one is trying to calculate with units in an orderly fashion; see Notes and Answers, note 1.3A.) It would also be obvious if the mistake were made of dividing instead of multiplying – since the 'ml' would not then cancel. If unsure whether to multiply the two quantities together, or to divide one by the other, one would only have to try out the three possible calculations to see which one yields a combination of units appropriate to excretion rate, i.e. mmol/h and not, say, ml²/(mmol h).

The calculation of **rates of substance flow** from products of concentration and fluid flow in that way is commonplace in physiology and the idea leads directly to the concept of **renal clearance**, and specifically to the use of inulin clearance as a measure of glomerular filtration rate (GFR). Often, when I have questioned students about inulin clearance, they have been quick to quote an appropriate formula, but have been unable to suggest appropriate units for what it yields. It is the analysis of the formula in terms of units that is my ultimate concern here, but a few lines on its background and derivation may be appropriate too. For the measurement of GFR, the plant polysaccharide inulin is infused into the body and measurements are later made of the concentrations in the blood plasma (P) and urine (U) and of the rate of urine flow (V). The method depends on two facts: first, that the concentration in the glomerular filtrate is essentially the same as the concentration in the plasma and, second, that the amount of inulin excreted is equal to the amount filtered. The rate of excretion is UV (as for urea) and the rate of filtration is GFR $\times$ P (again a flow times a concentration). Thus:

$$\text{GFR} \times P = UV,$$

so that:

$$\text{GFR} = \frac{UV}{P}. \tag{1.1}$$

Although the quantity calculated here is the GFR, it can also be thought of as the rate at which plasma would need to be completely cleared of inulin to explain the excretion rate (whereas in fact a larger volume is partially cleared). Hence the term 'renal plasma clearance'. The formula may be generalized to calculate clearances for other excreted substances:

$$\text{renal plasma clearance} = \frac{UV}{P}. \tag{1.2}$$

It may be obvious that GFR needs to be expressed in terms of a volume per unit time, but for the more abstruse concept of clearance the appropriate units are less apparent. This brings us to my main point, that appropriate units can be found by analysis of the formula.

If the concentrations are expressed as g/ml, and the urine flow rate is expressed as ml/min, then the equation can be written in terms of these units as follows:

$$\text{units for clearance} = \frac{\text{g/ml} \times \text{ml/min}}{\text{g/ml}}.$$

Since 'g/ml' appears on the top and bottom lines, it can be cancelled, leaving the right-hand side of the equation as 'ml/min'. Such units (volume per unit time) are as appropriate to clearances in general as to GFR.

To reinforce points made earlier, suppose now that equation 1.1 is wrongly remembered, or that the concentrations of inulin in the two fluids are expressed differently, say one as g/l and one as g/ml. If the calculation is written out with units, as advocated, then error is averted.

It has been emphasized that rates of substance flow can be calculated as products of concentration and fluid flow. In another context, the rate of oxygen flow in blood may be calculated as the product of blood oxygen content and blood flow, and the rate of carbon dioxide loss from the body may be calculated as the product of the concentration (or percentage) of the gas in expired air and the respiratory minute volume. Such ideas lead straight to the **Fick Principle** as applied, for example, to the estimation of cardiac output from measurements of whole-body oxygen consumption and concentrations of oxygen in arterial and mixed-venous blood. The assumption is that the oxygen consumption is equal to the difference between the rates at which oxygen flows to, and away from, the tissues:

oxygen consumption
= cardiac output $\times$ arterial $[O_2]$ − cardiac output $\times$ mixed-venous $[O_2]$
= cardiac output $\times$ (arterial $[O_2]$ − mixed-venous $[O_2]$),

where the square brackets indicate concentrations. From this is derived the Fick Principle formula:

$$\text{cardiac output} = \frac{\text{oxygen consumption}}{\text{arterial } [O_2] - \text{mixed-venous } [O_2]}. \tag{1.3}$$

Re-expressed in terms of units, this becomes:

$$\text{cardiac output} = \frac{\text{ml } O_2/\text{min}}{\text{ml } O_2/\text{l blood}} = \frac{\text{ml } O_2}{\text{min}} \times \frac{\text{l blood}}{\text{ml } O_2} = \text{l blood/min.}$$

Note two points. First, mistakes may be avoided if the substances (oxygen and blood) are specified in association with the units ('ml O_2/l blood' rather than 'ml/l'). Second, the two items in the bottom line of equation 1.3 have the same units and are lumped together in the treatment of units. Actually, since one is subtracted from the other, it is a necessity that they share the same units. Indeed, if one finds oneself trying to add or subtract quantities with different units, then one should be forced to recognize that the calculation is going astray.

We turn now to the **mechanical work** that is done when an object is lifted and when blood is pumped. When a force acts over a distance, the mechanical work done is equal to the product of force and distance. Force may be expressed in newtons and distance in metres. Therefore, work may be expressed in N m, the product of the two, but also in joules, since 1 J = 1 N m (Table 1.1). Conversion to calories, etc. is also possible, but the main point here is something else. When an object is lifted, the work is done against gravity, the force being equal (and opposite) to the object's weight. Weights are commonly expressed as 'g' or 'kg', but these are actually measures of mass and not of force, whereas the word 'weight' should strictly be used for the downward force produced by gravity acting on mass. A mass of 1 kg may be more properly spoken of as having a weight of 1 kg-force. Weight depends on the strength of gravity, the latter being expressed in terms of g, the gravitational acceleration. This is less on the Moon than here, and it is variable on the Earth in the third significant figure, but for the purpose of defining 'kg-force' the value used is 9.807 m/s², with 1 kg-force being 9.807 N (Table 1.1). This distinction between mass and weight is essential to the procedures advocated here for analysing equations in terms of units and including units in calculations to avoid error.

In relation to the pumping of blood, the required relationship is not 'work equals force times distance', but 'work equals increase in pressure times volume pumped'. If unsure of the latter relationship, can one check that it makes sense in terms of units? The analysis needs to be in terms of SI units, not, say, calories, mmHg and litres. Areas are expressed as m², and volumes as m³. Accordingly:

$$\text{work (J)} = \text{pressure} \times \text{volume} = \text{N/m}^2 \times \text{m}^3 = \frac{\text{N}}{\text{m}^2} \times \text{m}^3 = \text{N m} = \text{J}.$$

Next we have a situation requiring the definition of the newton as 1 kg m/s². The **pressure due to a head of fluid**, e.g. in blood at the bottom of a vertical blood vessel, is calculated as ρgh, where ρ is the density of the fluid, g is the gravitational acceleration (9.807 m/s²) and h is the height of fluid. To check that this expression really yields units of pressure (N/m²), we write:

$$\rho gh = \frac{\text{kg}}{\text{m}^3} \times \frac{\text{m}}{\text{s}^2} \times \text{m} = \frac{\text{kg}}{\text{m s}^2}$$

Recalling that 1 N = 1 kg m/s², we now write:

$$\text{pressure} = \frac{\text{N}}{\text{m}^2} = \frac{\text{kg m}}{\text{s}^2} \times \frac{1}{\text{m}^2} = \frac{\text{kg}}{\text{m s}^2},$$

which is the same expression as before.

There are some quantities for which the units are not particularly memorable for most of us, including peripheral resistance and the solubility coefficients for gases in liquids. Appropriate units may be found by analysis of the equations in which they occur. Peripheral resistance is discussed in Section 4.3, while here we consider the case of gas **solubility coefficients**, and specifically the solubility coefficient of oxygen in body fluids such as blood plasma. The concentration of oxygen in simple solution, $[O_2]$, increases with the partial pressure, P_{O_2}, and with the solubility coefficient, S_{O_2}:

$$[O_2] = S_{O_2} P_{O_2}. \tag{1.4}$$

The concentration may be wanted in ml O_2/l fluid or in mmol/l, with the partial pressure being specified in mmHg, kPa or atmospheres, but let us choose mmol/l and mmHg. Rearranging equation 1.4 we see that S_{O_2} equals the ratio $[O_2]/P_{O_2}$, so that the compatible solubility coefficient is found by writing:

$$\frac{[O_2]}{P_{O_2}} = \frac{mmol}{l} \times \frac{1}{mmHg} = \frac{mmol/l}{mmHg} = \text{mmol/l per mmHg or mmol/l mmHg.}$$

To reinforce the theme of how to avoid errors, note what happens if an incompatible form of solubility coefficient is used in a calculation. In different reference works, solubility coefficients may be found in such forms as 'ml/l per atmosphere', 'mmol/(l Pa)', etc., as well as mmol/l per mmHg. If the first of these versions were to be used in a calculation together with a gas pressure expressed in mmHg, then the units of concentration would work out as:

$$\frac{\text{ml } O_2/\text{l fluid}}{\text{atmosphere}} \times mmHg = \text{ml } O_2 \text{ mmHg}/(\text{l fluid atmosphere}).$$

The need to think again would at once be apparent.

The above illustrations have variously involved SI and non-SI units in accordance with need and convenience, but other methods of analysis are sometimes appropriate that are less specific about units, at least in the early stages. It is mainly to avoid complicating this chapter that a description of 'dimensional analysis' is consigned to Notes and Answers, note 1.3B, but it is also less generally useful than unit analysis. We look next at diffusion to illustrate a slightly different approach in which the choice of units is deferred.

Suppose that an (uncharged) substance S diffuses from region 1 to region 2 along a diffusion distance d and through a cross-sectional area a. The (uniform) concentrations of S in the two regions are respectively $[S]_1$ and $[S]_2$. The **rate of diffusion** is given by the following equation:

$$\text{rate} = ([S]_1 - [S]_2) \times a/d \times D, \tag{1.5}$$

where D is the 'diffusion coefficient'. The appropriate units for D may be found by rearranging the equation and proceeding as follows:

$$D = \frac{\text{rate}}{[S]_1 - [S]_2} \times \frac{d}{a} = \frac{\text{rate}}{\text{concentrations}} \times \frac{\text{distance}}{\text{area}}.$$

The rate of diffusion is the amount of S diffusing per unit of time and concentrations are amounts of S per unit of volume. Therefore:

$$D = \frac{\text{amount}}{\text{time}} \times \frac{\text{volume}}{\text{amount}} \times \frac{\text{distance}}{\text{area}} .$$

Following the practice adopted above, the various items in the right-hand expression could have been given in terms of kg, s, m³, m² and m, and that approach would be valid. Diffusion coefficients are in fact commonly given as cm²/s, so let us now specify distance, area and volume in terms of cm, cm² and cm³, and time in seconds. Then the expression becomes:

$$\text{units for } D = \frac{\text{amount}}{\text{s}} \times \frac{\text{cm}^3}{\text{amount}} \times \frac{\text{cm}}{\text{cm}^2} = \frac{\text{cm}^2}{\text{s}} .$$

Note that it is irrelevant here how the amount of substance is expressed, whether it be in g, mmol, etc. For another form of diffusion coefficient, relating to gas partial pressures, see Notes and Answers, note 1.3C.

It must be acknowledged finally that some equations are not sensibly analysed in terms of units. These are empirically derived formulae that have no established theoretical basis. For example, there are formulae that relate vital capacity, in litres, to age in years and body height in centimetres; there is no way of combining units of time and length to obtain units of volume. One must remember this general point to avoid being puzzled sometimes, but it is also true that the analysis of an empirical equation in terms of units or dimensions can sometimes lead to its refinement and to theoretical understanding.

Conclusions

Although the main theme here is the avoidance of error by consideration of units, it has also provided a context in which to introduce various commonly used formulae. In case these have obscured the ideas pertinent to the main theme, it may be helpful to summarize those ideas here.

1. Units can be combined, manipulated and cancelled like algebraic symbols.
2. The two sides of an equation must balance in terms of units as well as numerically.
3. If a formula calls for quantities to be expressed in particular units, then mistakes in this regard are preventable by writing them out as part of the calculation.

4. When quantities of more than one substance are involved, it is usually advisable to specify these along with the units, writing, for example, 'ml O_2/ml blood' rather than simply 'ml/ml' (which cancels, unhelpfully, to 1).

5. Quantities expressed in differing units cannot be combined by addition or subtraction.

6. Attention to units may prevent quantities from being inappropriately combined in other ways too (multiplied instead of divided, for example). Indeed it may suggest the right way of calculating something when other forms of reasoning falter.

7. Analysis of units may provide a partial check on half-remembered formulae.

8. Appropriate units for unfamiliar quantities can be found by analysing the equations in which they occur.

9. Weight (force) must be distinguished from mass (quantity).

10. Analysis of units sometimes requires knowledge that $1\,N = 1\,kg\,m/s^2$.

11. Units on the two sides of an equation may not balance if the relationship is empirical and has no theoretical basis.

To these ideas may be added two others, relating to indices and logarithms, that emerge in the next Section.

12. Exponents (indices) must be dimensionless, i.e. they can have no units.

13. Strictly it is not possible to take the logarithm of a number that has dimensions or units, although there are situations in which it is acceptable to do so.

Practice in unit analysis

Readers wishing to practise unit analysis might like to try the following exercises (some relating to physics rather than physiology). Help is given Notes and Answers.

1. If SI units for viscosity are unfamiliar, find them by analysing **Poiseuille's equation**. This relates the rate of flow of fluid, i.e. volume per unit time, in a cylindrical tube (e.g. blood in a blood vessel) to viscosity, to the radius and length of the tube and to the difference in hydrostatic pressure between its two ends:

$$\text{flow rate} \propto \text{pressure difference} \times \frac{\text{radius}^4}{\text{viscosity} \times \text{length}}. \tag{1.6}$$

2. Einstein's '$E = mc^2$' is well known. Treating energy, mass and velocity in terms of SI units, show that the two sides of the equation are compatible.

3. If 'RT/zF' is already familiar in relation to the Nernst equation, analyse it in terms of units. Its components are given in Appendix A, while the units for the whole expression are 'volts'. For this exercise, use the versions of R and F that involve calories. Appendix A also gives F in terms of coulombs; I have seen it given in physics textbooks as 'coulombs', 'coulombs/equivalent' and 'coulombs/volt equivalent', and this suggests another exercise. I give F as coulombs/volt equivalent, but is that correct? More specifically, do the relationships discussed in Section 7.6 then work out correctly in terms of units?

4. If the formula for calculating the period of a simple pendulum was once known, but is now forgotten, try reconstructing it by unit analysis, albeit partially, given only that the period increases with pendulum length and decreases with g.

1.4 Analysis of units in expressions involving exponents (indices)

Two main points are made here in relation to the unit analysis of equations containing exponents, one concerning the exponents themselves and the other having to do with other constants. At the same time, the opportunity is taken to say a little about exponential time courses and allometric relationships. The basic rules for working with exponents (indices) are given in Appendix B.

The first point is simply that exponents must be dimensionless quantities; they cannot have units. Thus, '3^2 eggs' is meaningful, but '$3^{2\ \text{eggs}}$' is not. While the 2 in 3^2 eggs is a simple number, exponents can also be expressions containing two or more variables that do have units – such as $3^{a/b}$, for example. This is satisfactory provided that the units cancel out. Thus, $3^{(4\ \text{eggs}/2\ \text{eggs})}$ equals 3^2. As a more serious example, and one commonly encountered in physiology, the simplest kinds of exponential time course are described by equations of the form:

$$Y = Y_0\, e^{kt}, \tag{1.7}$$

where Y is the variable in question, t is time (in seconds, say), Y_0 is the the value of Y when $t = 0$ and k is a constant (the 'rate constant', often negative) with units of time^{-1} (here s^{-1} or $1/s$). The e has its usual meaning, a number close to 2.718. Here the units in kt cancel out (i.e. s$/$s = 1). An alternative to e^{kt} in equation 1.7 is $e^{t/\tau}$ where the commonly used symbol τ (tau) is equal to $1/k$, and is called the 'time constant'. This has the same units as t, so that t/τ, like kt, is dimensionless.

The second point is one that could be harmlessly ignored (as it is by many physiologists) were it not that I have put so much emphasis on unit analysis. It concerns certain kinds of empirical relationship, as opposed to relationships founded in theory. Countless physiological and anatomical measurements have been made on mammals of different sizes, from shrews to whales, and the relationships between these and body mass have been explored. (In relation to purely human physiology, one may likewise explore relationships in individuals of differing size.) In very many cases the variable, Y, has been found to depend on body mass, M, in accordance with this equation:

$$Y = a M^b, \tag{1.8}$$

where a and b are constants. There is always some statistical scatter in these so-called 'allometric' relationships, with consequent uncertainty about the best values of the constants. To start with a case that gives no problem with unit analysis, it appears that heart mass is near-enough exactly proportional to M over seven orders of magnitude, such that $Y = 0.006 M^{1.0}$, with both masses in kg. (This implies that the heart makes up about 0.6% of body mass over the full size range.) There is no difficulty with units here, the '0.006' having none. To see how problems can arise, consider next the case of skeletal mass.

As Galileo pointed out in 1637, relative skeletal mass should increase with body mass, at least in land mammals, if the largest are not to collapse under their own weight (or the smallest are not to be burdened with extra bone). Here is an equation that has been fitted to data on dry skeletal mass (Prange *et al.*, 1979):

$$\text{skeletal mass (kg)} = 0.061\, M^{1.09}. \tag{1.9}$$

Now there is a difficulty, for $M^{1.09}$ has units of kg$^{1.09}$ and this suggests that the '0.061' has units of kg$^{-0.09}$ (with some uncertainty due to scatter in the data). This makes no obvious sense. A solution is to divide M by some reference mass, most conveniently 1 kg, so that the equation becomes, in the latter case:

$$\text{skeletal mass (kg)} = 0.061 \left[\frac{M}{1 \text{ kg}}\right]^{1.09}. \tag{1.10}$$

Unlike M, the ratio $M/(1 \text{ kg})$ is dimensionless. On this basis, the '0.061' is in kg, like skeletal mass. Put more generally, the constant a in equation 1.8 comes to have the same units as Y. Usually this rather pedantic procedure is not explicitly followed and no harm results. There is more on allometric relationships in Sections 1.5, 3.10, 3.12 and 6.16.

1.5 Logarithms

Physiologists use logarithms in a variety of contexts, notably in relation to membrane potentials (Nernst equation), acid–base balance (pH, Henderson–Hasselbalch equation), sensory physiology (Weber–Fechner 'law') and graphical analysis (of exponential time courses, allometry, dose-response curves). Since logarithms now play a much smaller part in school mathematics than formerly, they are explained in Appendix B. The main purpose of this Section is to say a little more about their use in the contexts just mentioned, but it concludes by returning briefly to the topic of rough calculation. Given the emphasis I have placed on unit analysis earlier in the chapter, I must first make a comment relating to that.

On the matter of units, it should be noted that strictly one can only take logarithms of dimensionless numbers, i.e. quantities that lack units. I say 'strictly' because people do commonly flout this rule, and do so without consequent difficulties or opprobrium. Thus, the elementary, and oldest, definition of pH is that it equals $-\log_{10}[H^+]$, where $[H^+]$ is the concentration of hydrogen ions in mol/l, the units being simply ignored in the calculation. The definition is in fact an oversimplification (Section 8.1), but we can move just one step towards a better definition by dividing $[H^+]$ by a standard concentration, $[H^+]_s$, of 1 mol/l {so that pH is defined as $-\log_{10}([H^+]/[H^+]_s)$}. The units of concentration are thus removed, while the number is unaffected (see the treatment of indices in Section 1.4). This exemplifies a general solution to the problem of taking logarithms of a quantity that has units: instead of ignoring them, one divides the quantity by some reference value, usually with a numerical value of 1. The next paragraph refers to logarithms of certain quantities Y and M; for propriety, these may be regarded as each divided by a reference quantity of one unit.

One use for logarithms is in the graphical analysis of exponential and allometric relationships (equations 1.7 and 1.8). In the case of equation 1.7, a graph of $\ln Y$ ($= \log_e Y$) against t yields a straight line of gradient k. Alternatively, a graph of $\log_{10} Y$ against t gives a straight line of gradient $k \log_{10} e$. An example is shown in Figure 5.4. In the case of equation 1.8, a graph of $\log Y$ against $\log M$ yields a straight line of gradient b.

Actually, there is sometimes another reason for plotting logarithms in these contexts. This is notably true in relation to the allometry of mammals of widely varying size for, on a linear scale, it is simply too hard to show comfortably the masses of shrews, whales, and all mammals in between. To cope with that great range of masses, one may plot $\log M$ (ignoring the mass units to do so), or else show actual values of M, using a logarithmic scale (e.g. showing, say, 0.1 kg, 1 kg, 10 kg, etc. at equally spaced intervals). Logarithmic scales are often used, at least partly for the same reason, for displaying drug concentrations (for dose-response curves).

Returning to the subject of hydrogen ion concentrations, these too vary over a huge range of magnitudes, and this is one reason why people prefer to work with pH. Thus, 10^{-4} and 10^{-8} mol/l water translate to pH 4 and pH 8 respectively. Sound intensities likewise vary enormously, making the logarithmic decibel scale convenient for the same reason. The decibel scale ties in with the Weber–Fechner law, the tendency for sensation to vary (not always exactly) with the logarithm of stimulus intensity.

In line with the logarithmic nature of pH, the Henderson–Hasselbalch equation, relating pH to P_{CO_2} and bicarbonate concentration, is usually formulated in logarithmic terms (see Notes and Answers):

$$pH = pK_1' + \log \frac{[HCO_3]}{[CO_2]}, \tag{1.11}$$

where pK_1' is a dissociation constant. (It will be apparent soon why the equation is expressed this way, rather than more usefully as in equation 8.1, where SP_{CO_2} replaces $[CO_2]$.)

The pH meter responds linearly to $\log [H^+]$ (as if it were a sense organ obeying the Weber–Fechner law). This is because the electrical potential across the glass membrane of the electrode, E_H, depends, at equilibrium, on the hydrogen concentrations (or rather activities) on its two sides in accordance with the Nernst equation:

$$E_H = \frac{RT}{zF} \ln \frac{[H^+]_1}{[H^+]_2}. \tag{1.12}$$

(R, T and F are often described simply as 'having their usual values'; they are given in Appendix A. z is the valency of the hydrogen ion, i.e. 1.) The subscripts 1 and 2 denote the two sides of the membrane. The pH on the inside of the glass electrode is constant. This description of the pH electrode is incidental, but the Nernst equation is essential to the understanding of cell membrane potentials and ion transport, and it is in these contexts that the equation is more often encountered. Here it is reformulated for the equilibrium potential of potassium (at 37 °C):

$$E_K \text{ (mV)} = 61.5 \log \frac{[K^+]_1}{[K^+]_2}. \tag{1.13}$$

Note that we have here the logarithm of a ratio, the ratio of two quantities expressed in identical units, i.e. $[K^+]_1$ and $[K^+]_2$. The same is true of equations 1.11 and 1.12. Such ratios are dimensionless, so that there is no problem here of taking the logarithms of quantities that have units. However, a further point can be made in this connection. Note that the expression log $([K^+]_1/[K^+]_2)$ is equal to (log $[K^+]_1$ − log $[K^+]_2$); if the first is valid, so too is the latter. Where there is a difference between two logarithms like that, the impropriety of one is cancelled out by the impropriety of the other.

Finally, we return to the subject of approximate arithmetic. In Appendix B there is a brief comment on the effects on calculations of inaccuracies occurring in logarithmic terms. (Question: how wrong might $[H^+]$ be if pH is only accurate to two decimal places?) Appendix B also emphasizes the usefulness of remembering that $\log_{10} 2$ is close to 0.30. Let us explore an example. Equations 1.11, 1.12 and 1.13 each include the logarithm of a concentration ratio. If this ratio starts with a value A, and then doubles to $2A$, then the logarithm of the ratio increases by 0.30 (because $\log 2A = \log 2 + \log A$). Likewise, halving the ratio decreases its logarithm by 0.30. With the Henderson–Hasselbalch equation in mind, we can therefore see, without further calculation, that doubling of $[HCO_3^-]$ or halving of $[CO_2]$ should raise the pH by 0.30. Since $[CO_2]$ is proportional to P_{CO_2}, it is also true that halving P_{CO_2} would raise the pH by 0.30. Let us put this into the context of an approximate calculation that does not even require the back of an envelope:

Question: At constant P_{CO_2}, could a rise in bicarbonate concentration from 20 mM to 30 mM explain a rise in pH from 7.10 to 7.43?

Answer: No – even a doubling of concentration only leads to a rise of 0.3 unit.

2 Quantifying the body: interrelationships amongst 'representative' or 'textbook' quantities

In order to develop a quantitative understanding of the workings of the body it is helpful to have at one's fingertips some representative, 'standard' or 'textbook' physiological quantities. Indeed, it is commonplace that students be expected to learn such figures for blood volume, cardiac output, respiratory minute volume, glomerular filtration rate, etc. These may well be those of a 'standard 70-kg man' or someone rather like him – young, healthy, not too fat, male and, as such, never pregnant or lactating (see Notes and Answers, note 2A). The purpose of this chapter is to illustrate how certain representative quantities may be made meaningful and so worth memorizing. Whether students try to learn every such quantity brought to their attention or as few as possible (neither extreme being desirable), the effort is most rewarding if those quantities are made meaningful in particular contexts and integrated into a general quantitative picture of bodily function. At this stage readers might care to list all those physiological quantities that they either know or feel they ought to know, and to specify contexts in which they think the knowledge would illuminate discussion of physiological topics.

One difficulty for learning is that physiological quantities vary from person to person and from moment to moment. At the same time, many people find it easiest to learn definite numbers or definite normal limits, finding the ill-defined ranges of reality too slippery to stay in the mind. Whilst the arithmetic of this book is mostly conducted with particular numbers, physiological variablity must never be forgotten. A couple of paragraphs on this is therefore apposite.

Core body temperature, like the concentrations of some solutes in extracellular fluid, is normally well regulated and for many purposes can be characterized by a single figure, 37 °C, despite the small, but significant, diurnal fluctuations and variation through the menstrual cycle. (The normal variation is little more than ± 1 °C, or $\pm 0.3\%$ in terms of the Kelvin scale.) In contrast to core body temperature, heart rate varies greatly, not just from person to person, but in single individuals according to such things as physical activity and mental state. Nevertheless, so long as this variability is appreciated, it is reasonable to have in mind that the average heart rate of a resting person is, as some textbooks say, about 72 beats/min. At the same time, there is no need for the concern felt by some students that other books give, say, 70 or 75 beats/min instead.

Some physiological variables must obviously vary with body size, e.g. basal metabolic rate and cardiac output, and it has long been the practice to quote these for a 'typical person', whether or not the latter is defined. The concept of the 'standard 70-kg man' is valuable in that it excludes variations related to body size, fatness, etc. (see Notes and Answers, note 2A). Whether or not 70 kg is near-average for a particular population, as it once was for young American adult males (and specifically the medical students who volunteered for measurement), depends largely on its current state of nutrition, and in some societies the recent trend has been to grow both taller and fatter.

Aside from this matter of variability, a major point to be made in this chapter is that numerical quantities generally only acquire meaning when they are compared in some way to other numerical quantities. Often the context of a comparison is one of change and the recognition of change or abnormality, as when blood pressure rises in exercise and disease. The colloid osmotic pressure of the blood plasma acquires meaning by comparison with blood pressures in capillaries, notably including those of the renal glomeruli and lung alveoli. Our concern here is with various physiological values that are commonly given in elementary textbooks and physiology courses, but more specifically with showing how groups of them can be interrelated through simple arithmetic. The items so treated are highlighted below in italics. Typical textbook values for individuals at rest are used where possible, with preference given to round numbers and to those that are most mutually compatible. Readers are free to substitute other values, and might even benefit from working through the calculations with them. Figure 2.1 maps the interrelationships amongst most of the physiological variables that are treated in this chapter. The end-product of all the calculations can best be described not as 'standard' (since that word has inappropriate connotations), but as a 'typical textbook man', with 'typical' referring as much to textbooks as to men.

Heart rate, stroke volume and *cardiac output* are interdependent in that the product of the first two yields the third. This checks with sufficient accuracy using respective values for a resting person of, say, 72 beats/min,

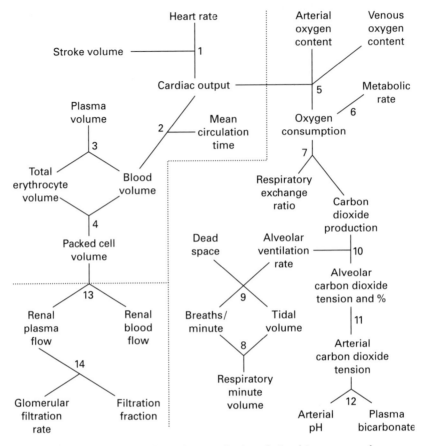

Fig. 2.1. The text explores the quantitative relationships amongst the
physiological variables shown here. Where two to four items are connected
by lines, the value of any one may be calculated from the value(s) of the
other(s). Calculations are illustrated in the text in the order shown by the
numbers, starting with the interdependence of heart rate, stroke volume
and cardiac output.

70 ml (= 0.07 l) and 5 l/min, for 72/min × 0.07 l = 5.04 l/min. The resting
cardiac output happens (in ourselves, but not in most mammals) to be
numerically the same as the typical *blood volume* inasmuch as the latter is
commonly given as 5 l. This implies that the *mean circulation time* (at rest),
namely the time for blood to pass once through either the systemic or the
pulmonary circulation, is 1 min. (An impressive figure to compare this with

is the circulation time in a shrew. With a resting heart rate of about 10 beats per second, the shrew's circulation time is only about 3 s – reducing to about 1 s in exercise.)

The human blood volume of 5 l consists of a *plasma volume* of 3 l and a *total erythrocyte volume* of 2 l (plus a tiny volume of white blood cells). The ratio of erythrocyte volume to total blood volume, i.e. 0.4, is known as the '*packed cell volume*' or '*haematocrit*' and is often expressed as a percentage, here 40%. This is slightly low compared with typical textbook averages (e.g. 46% in men and 41% in women), but the discrepancy is explained in Section 4.1.

The resting cardiac output of 5 l blood/min relates to oxygen carriage. Arterial blood contains 200 ml O_2/l blood (*arterial oxygen content*). The corresponding *oxygen content of mixed-venous blood* in a resting person is 150 ml O_2/l blood. Therefore oxygen is transported in the blood from the left ventricle at a rate of:

$$5 \text{ l blood/min} \times 200 \text{ ml } O_2/\text{l blood} = 1000 \text{ ml } O_2/\text{min}.$$

It is transported back to the right atrium at a rate of:

$$5 \text{ l blood/min} \times 150 \text{ ml } O_2/\text{l blood} = 750 \text{ ml } O_2/\text{min}.$$

The difference between these two rates, i.e. $1000 - 750 = 250$ ml O_2/min, is the resting *rate of oxygen consumption*. Effectively, though not explicitly, this calculation uses the Fick Principle formula (equation 1.3).

The rate of oxygen consumption is related to energy consumption (metabolic rate), with the consumption of each litre of oxygen being associated with about 4.8 kcal (20.1 kJ), as discussed in Section 3.2. The resting oxygen consumption of 250 ml O_2/min thus corresponds to a *resting metabolic rate* of:

$$0.25 \text{ l } O_2/\text{min} \times 4.8 \text{ kcal/l } O_2 = 1.2 \text{ kcal/min}.$$

This is the same as 5.0 kJ/min, 72 kcal/h (301 kJ/h) and 1728 kcal/day (7233 kJ/day) – or, as round-number values, 70 kcal/h (300 kJ/h) and 1700 kcal/day (7000 kJ/day).

The resting rate of oxygen consumption may be used to calculate the corresponding rate of *carbon dioxide production* – by multiplying it by the *respiratory exchange ratio* (itself defined as the rate of carbon dioxide production divided by the rate of oxygen consumption). Although the respiratory exchange ratio varies with the fuels being metabolized, it is typically near 0.8 (Table 3.1). Although it is not normally quoted with units, for the next calculation it may be helpful, in the spirit of Chapter 1, to think of them as '(ml CO_2/min)/(ml O_2/min)'. Accordingly, the resting rate of carbon dioxide production is:

$$0.8 \text{ (ml } CO_2/\text{min)}/(\text{ml } O_2/\text{min}) \times 250 \text{ ml } O_2/\text{min} = 200 \text{ ml } CO_2/\text{min}.$$

The respiratory exchange ratio may also be used in calculating the increase in carbon dioxide content between arterial and mixed-venous blood from the corresponding decrease in oxygen content. On the basis of figures already given (and with units for respiratory exchange ratio omitted this time), the increase in carbon dioxide content is:

$$0.8 \times (200 \text{ ml } O_2/\text{l blood} - 150 \text{ ml } O_2/\text{l blood}) = 40 \text{ ml } CO_2/\text{l blood}.$$

This is consistent with an increase from 480 ml CO_2/l in arterial blood to 520 ml CO_2/l in mixed-venous blood, figures that might be useful in sketching a carbon dioxide dissociation curve.

Just as cardiac output is the product of heart rate and stroke volume, so *respiratory minute volume, pulmonary ventilation rate* or *total (pulmonary) ventilation* is the product of *breathing frequency* and *tidal volume* (volume of each breath). For the resting state, a tidal volume of 500 ml may be combined with a breathing rate of, say, 11 breaths per minute, which is about average for men and women. Then the respiratory minute volume is:

$$11/\text{min} \times 500 \text{ ml} = 5500 \text{ ml/min}.$$

The rate of *alveolar ventilation* is less than this because of the *dead space,* namely the volume of the respiratory passages between the external environment and the alveoli. If this is taken as 155 ml, the resting rate of alveolar ventilation is:

$$11/\text{min} \times (500 \text{ ml} - 155 \text{ ml}) = 3795 \text{ ml/min}.$$

Actually, I have seen neither the 11/min nor the 155 ml figures given as such as representative in textbooks, but only figures that are higher and lower than these; they are chosen to lead us exactly to our destination in the next paragraph. As an approximation, the dead space is said to be numerically equal to the body mass in pounds: 70 kg is 154 lb, and that happens to be in fair accord with the chosen dead space of 155 ml. As a representative value, 3795 ml/min is far too precise, and either 3.8 or 4 l/min would be more sensibly remembered.

If a person is breathing out alveolar air at 3795 ml/min while carbon dioxide is expired as part of it at a rate of 200 ml/min, as above, then the *percentage of carbon dioxide in the alveolar air* is:

$$100 \times (200 \text{ ml/min})/(3795 \text{ ml/min}) = 5.27\%.$$

More memorable than a percentage, however, is the corresponding tension of carbon dioxide. If the atmospheric pressure is taken as 760 mmHg, the *alveolar carbon dioxide tension* is 5.27% of this, i.e. 40 mmHg. In practice, the logic of these two paragraphs may be reversed in order to estimate dead space, the carbon dioxide content of alveolar air being measured directly from a sample collected at the end of expiration. (The volumes in these two paragraphs may be taken as being for gas saturated with water vapour at body temperature. However, the rate of CO_2 production and the percentage of CO_2 in the alveoli are more usually expressed in terms of dry gas at standard temperature and pressure.)

The alveolar carbon dioxide tension is virtually the same as the *arterial carbon dioxide tension*, and this is a determinant of *plasma pH*. More exactly, pH is determined by the ratio of *plasma bicarbonate concentration* to carbon dioxide tension – in accordance with the Henderson–Hasselbalch equation (Section 8.2). For a normal arterial pH of 7.4 and a normal arterial P_{CO_2} of 40 mmHg, the bicarbonate concentration is 24 mmol/l.

Turning now to kidney function, the *renal blood flow* is about 1.2 l/min. It makes up just under a quarter of the cardiac output if that is taken as 5 l/min. Since 60% of the blood is plasma, according to the haematocrit value given earlier, the *renal plasma flow* is 60% of the renal blood flow, namely 0.6×1.2 l/min = 0.72 l/min. However, given the variability of all these quantities both between and within individuals (and the fact that the 40% haematocrit is low for a male), the latter figure is better rounded to 0.7 l/min. The high blood flow is not necessary to supply the kidneys with oxygen, but is required for the high *glomerular filtration rate*. This is often given for men as 125 ml/min (an average figure exactly equivalent to 180 l/day). The *filtration fraction* is the ratio of glomerular filtration rate to renal plasma flow. Its value here is equal to (0.125 l/min)/(0.7 l/min), i.e 0.18, or 0.2 as a round number. This is the final item to be shown in Figure 2.1.

The total solute concentration of plasma and other body fluids is commonly expressed in terms of *milliosmolality*, a measure related to osmotic pressure and to the millimolal concentrations of all the various solutes present (see Section 7.10 and the associated note). A round-number value of 300 mosmol/kg water is commonly used, for example, in descriptions of the renal countercurrent mechanism. Of all the plasma solutes contributing to this figure, the most abundant is sodium, and, because each cation is balanced by an anion (principle of electroneutrality – Section 7.6), the sodium concentration is not far from $300/2 = 150$ mmol/kg water. Translated into the more usual units of mmol/l plasma (mM), this figure for *plasma sodium* is slightly under 150 mmol/l (Section 7.9). In cells, the principle cation is not sodium, but potassium; *cell potassium* varies in concentration, but is typically close to the same figure of 150 mmol/kg water.

The 3 l of plasma is part of the *extracellular fluid volume* of 14 l that also includes the interstitial fluid. (Extracellular fluid may also be defined as including the lesser amounts of water in cerebrospinal fluid, bone, etc. – see Notes and Answers, note 2B.) The volume of water in the cells, the *intracellular volume*, is twice that, i.e. 28 l. If the total amount of water (42 l) is that in a 70-kg body, then the *water content of the body* is 60%.

From the volume of the extracellular fluid and the concentration of sodium within it may be estimated the *amount of sodium in extracellular fluid*:

$$14\,l \times 150 \text{ mmol/l} = 2100 \text{ mmol.}$$

This may not be a particularly memorable figure, but it can be compared with *dietary sodium intake* (roughly 80–200 mmol/day on a typical western diet) and the total *sodium content of the body* (about 3500 mmol in a 70-kg person).

The sequence of steps by which alveolar P_{CO_2} was calculated above is equivalent to applying the following formula, with the inspired P_{CO_2} being taken as near-enough zero:

$$\begin{matrix} \text{rate of carbon} \\ \text{dioxide} \\ \text{production} \end{matrix} = \begin{matrix} \text{rate of} \\ \text{alveolar} \\ \text{ventilation} \end{matrix} \times \frac{\text{alveolar } P_{CO_2} - \text{inspired } P_{CO_2}}{\text{atmospheric pressure}}. \quad (2.1)$$

In analogous manner we may also write:

$$\begin{matrix} \text{rate of} \\ \text{oxygen} \\ \text{consumption} \end{matrix} = \begin{matrix} \text{rate of} \\ \text{alveolar} \\ \text{ventilation} \end{matrix} \times \frac{\text{inspired } P_{O_2} - \text{alveolar } P_{O_2}}{\text{atmospheric pressure}}. \quad (2.2)$$

Unlike inspired P_{CO_2}, inspired P_{O_2} is very far from zero. To consider first dry atmospheric air, its P_{O_2} is equal to the product of the total atmospheric pressure, let us say 760 mmHg (standard atmospheric pressure), and the fraction of oxygen in the air, i.e. 0.21 (= 21%). Therefore the P_{O_2} of dry atmospheric air is:

$$0.21 \times 760 \text{ mmHg} = 160 \text{ mmHg}.$$

The P_{O_2} of inspired air reaching the alveoli is lower because of its moisture content. Inspired air that is fully moistened at 37 °C has a water vapour pressure of 47 mmHg, so that its P_{O_2} is:

$$0.21 \times (760 - 47) \text{ mmHg} = 150 \text{ mmHg}.$$

We now have values for all the terms in equation 2.2 except for alveolar P_{O_2}, so the latter may now be calculated from the others. The first stage is to rearrange the equation thus:

$$\begin{matrix} \text{alveolar} \\ P_{O_2} \end{matrix} = \begin{matrix} \text{inspired} \\ P_{O_2} \end{matrix} - \begin{matrix} \text{atmospheric} \\ \text{pressure} \end{matrix} \times \frac{\text{rate of oxygen consumption}}{\text{rate of alveolar ventilation}} . \quad (2.3)$$

This time, as in subsequent pages, I present the calculation as a 'problem' for the reader, but the result may be checked in 'Notes and Answers'.

....

2.1 Assuming that oxygen is consumed at 250 ml/min, that the rate of alveolar ventilation is 3795 ml/min, and that the inspired P_{O_2} is 150 mmHg, what is the alveolar P_{O_2}?

....

Finally, we may combine equations 2.1 and 2.2 in order to find a different relationship amongst some of the figures already encountered. When the body is in a steady state, the respiratory quotient is equal to the respiratory exchange ratio, namely the output of carbon dioxide divided by the intake of oxygen. This is equal to the right-hand side of equation 2.1 divided by the right-hand side of equation 2.2. Cancelling out 'rate of alveolar ventilation' and 'atmospheric pressure' from the top and bottom lines, we thus have:

$$\text{respiratory exchange ratio} = \frac{\text{alveolar } P_{CO_2} - \text{inspired } P_{CO_2}}{\text{inspired } P_{O_2} - \text{alveolar } P_{O_2}}$$

$$= \frac{\text{alveolar } P_{CO_2}}{\text{inspired } P_{O_2} - \text{alveolar } P_{O_2}} . \quad (2.4)$$

....

2.2 Is the respiratory exchange ratio calculated from equation 2.4 the same as the respiratory quotient given above (i.e. 0.8)?

....

In conclusion, the student wondering what quantities are worth learning should look for contexts in which each has meaning. Not only should thinking about quantitative relationships aid memorization, but remembering how to calculate one quantity from others reduces the number to be learnt.

3 Energy and metabolism

To feel at home in this field one must be familiar with a variety of measures and units of energy – calories, joules, litres of respired oxygen, slices of bread, and so on. The more books or papers one consults, the more evident this becomes. Many of the early calculations in this chapter are intended primarily to help with this problem. Fortunately, physiology makes little use of such other units as kilowatt-hours, British thermal units and litre-atmospheres. Amongst calculations on metabolic rate are some to do with its dependence, in mammals generally, on body size (Section 3.10). Uses and limitations of such allometric scaling are then explored in relation to drug dosage, life span, heart rate, the metabolic cost of sodium transport and the rate of metabolic production of water (Sections 3.11–3.14).

Most emphasis is placed here on aerobic metabolism, and on glucose rather than on other metabolic substrates. The reader might like to re-work some of the calculations in terms of, say, anaerobic metabolism or fatty acids.

Amounts of energy and work are expressed here in terms of both kilocalories and kilojoules. When pairs of figures (i.e. kcal and kJ) are chosen for their convenience, or as round numbers, then they are not always exactly equivalent. Calculations on energy use are also to be found in Sections 4.7, 6.8, 9.4 and 9.5.

3.1 Measures of energy

The variety of units in which energy and work may be expressed does not help casual quantitative thinking. Most of this chapter makes use of kilocalories (kcal) with kilojoules (kJ) in parentheses. As shown in Table 1.1, 1 kcal is about 4.19 kJ, so that 1 kJ is 0.24 kcal. When calculations in the two units lead here to slightly different answers, this is due to rounding errors.

By definition, it takes 1 kcal to heat 1 kg of water by 1 °C (strictly, from

14.5 °C to 15.5 °C). It takes about 0.8 kcal to raise the temperature of 1 kg of human tissue by the same amount. (In other words, the specific heat capacity of the body is 0.8 kcal/kg per °C.)

3.1.1 How many kilocalories are required to raise the temperature of a 100-kg person by one degree?

To put this answer into a context, suppose that the person in question 3.1.1 has a basal metabolic rate of 1900 kcal/day (8000 kJ/day). This is equivalent to nearly 80 kcal/h (333 kJ/h), enough to raise the temperature by about one degree per hour if there were no heat loss.

Another unit of energy is the metre kilogram-force, where 1 m kg-force is 9.81 J. (Although this may often be appropriately rounded up to 10 J, as already noted, it is as well not to lose sight of the fact that the 9.81, as m/s², represents g, the acceleration due to gravity.)

3.1.2 In accordance with the last paragraph, how much energy in (a) kJ and (b) kcal is required to raise a 100-kg person by 10 m?

The answers make no allowance for inefficiency, e.g. for heat generation within the body of the person doing the lifting (see Section 3.9).

In this context of work against gravity, joules have an obvious advantage over calories, and later calculations on jumping and lifting (Section 9.4) are framed in terms of joules. In quantifying the body's energy expenditure it is also commonplace to think in terms neither of joules nor calories, but of oxygen consumption. Accordingly, the relationship between energy and oxygen consumption is discussed next.

3.2 Energy in food and food reserves; relationships between energy and oxygen consumption

The metabolic oxidation of glucose may be described by the following overall reaction:

$$C_6H_{12}O_6 + 6 O_2 = 6 CO_2 + 6 H_2O. \qquad [3.1]$$

From this, it is evident that the complete oxidation of 1 mol of glucose (180 g) to carbon dioxide and water utilizes 6 mol of oxygen and yields 6 mol of

Table 3.1. *Energy content, in kcal/g and kJ/g, of carbohydrates and of typical fat and protein. Also shown are the amounts of oxygen required for complete combustion or metabolism of the carbohydrates and fat, the amounts of energy released per litre and per mole of oxygen, and the respiratory quotient (the ratio of carbon dioxide molecules released to oxygen molecules consumed). The equivalent figures for protein allow for the loss of energy in nitrogenous excretory products*

	Glucose	Glycogen, starch	Fat	Protein
kcal/g	3.7	4.2	9.3	4.4
kJ/g	15.6	17.5	39.1	18.5
l oxygen/g	0.75	0.83	1.98	0.96
kcal/l oxygen	5.0	5.0	4.7	4.6
kJ/l oxygen	20.9	21.1	19.8	19.2
kcal/mol oxygen	112	113	106	103
kJ/mol oxygen	468	473	444	430
Respiratory quotient	1.00	1.00	0.71	0.81

carbon dioxide. Also, since 1 mol of oxygen at standard temperature and pressure (STP) occupies 22.4 l, 1 mol of glucose requires 134.4 l of oxygen. Table 3.1 gives some approximate conversion factors for glucose in accordance with these figures – and corresponding factors for fat and protein. The energy data are as determined by combustion in a bomb calorimeter, except that those for protein allow for the energy content of the nitrogenous excretory products and are thus somewhat lower than are obtained by bomb calorimetry.

We store some energy in our bodies as glycogen and protein, but very much more as fat. What is the advantage of storing fat, apart from its ability to pad and insulate? The simple answer is that fat yields more than twice as much energy per gram when fully metabolized than does glycogen or protein. The difference between fat and glycogen is increased by the fact that glycogen in cells is necessarily associated with water, binding an amount roughly twice its own mass. Therefore, a gram of fat stores more than six times as much energy as does a gram of hydrated glycogen, i.e. $(9.3 \text{ kcal/g}) / (4.2 \text{ kcal/g}) \times (2 + 1) = 6.6$.

3.2.1 Consider a 70-kg man containing 10 kg fat. How much heavier would he have to be to store the same amount of energy in the form of glycogen?

Whole-body energy expenditure is commonly estimated from oxygen consumption. There is no unique conversion factor, since we metabolize carbohydrates, proteins and lipids in variable proportions. However, factors of about 4.8 kcal/l oxygen and 20 kJ/l oxygen have often been used, these being within the ranges for the four substances in Table 3.1 (4.6–5.0 kcal/l oxygen and 19–21 kJ/l oxygen). More accurate estimates employ measurements not just of oxygen consumption, but of carbon dioxide production and nitrogen excretion, since it is then possible to assess the relative proportions of carbohydrate, fat and protein involved (see Notes and Answers).

3.2.2 On the basis of Table 3.1, what is the (maximum) percentage error involved in assuming a conversion factor of 4.8 kcal/l oxygen if respiration were actually to involve only carbohydrate or only protein?

The relationship between heat production and respiratory exchanges was first explored by Lavoisier and Laplace in 1780, but they measured carbon dioxide production rather than oxygen consumption. They found then that the ratio of carbon dioxide to heat production (measured as the weight of water released from melting ice) was almost the same for guinea-pigs as for burning charcoal, and this led them to the important realization that respiration is a form of combustion. (The discrepancy corresponded to a respiratory exchange ratio of about 0.8, which we now know to be typical – see Chapter 2. Lavoisier and Laplace saw later that the discrepancy was due to the metabolism of hydrogen.)

3.3 Basal metabolic rate

The basal metabolic rate (BMR) of a 70-kg man is usually within the range of 1200–2100 kcal/day (5000–9000 kJ/day), depending on such factors as fat content and age. A representative figure of 1700 kcal/day (7000 kJ/day) was used in Chapter 2, this being equivalent to a resting oxygen consumption of about 250 ml/min. When the body is in energy balance, and not gaining or losing fat, these figures should correspond to the dietary energy intake, less faecal losses. It is not hard to visualize one's own daily food intake, but it includes water, the bubbles in bread and indigestible cellulose, and one's metabolic rate over a full day is generally well above basal. It may, therefore, be helpful to think of the above basal metabolic rate in terms of purer fuels.

3.3.1 For a metabolic rate of 1700 kcal/day (7000 kJ/day), what would the daily rates of fuel consumption be, in grams, if the fuel were (a) entirely starch and (b) entirely fat? (See Table 3.1.)

Metabolic rates may also be expressed in watts and it is therefore useful to know the approximate 'wattage' of a resting body. This also allows direct comparisons with familiar electrical appliances and, since most of the energy used by the body ultimately heats the environment (especially when little mechanical work is done), it can be revealing to compare, say, the total metabolic rate of an audience in a lecture theatre with the power of an electric fire (commonly 1–2 kW). However, a more useful first step is perhaps to compare a single individual with a light bulb.

3.3.2 If a person's metabolic rate happens to be 8640 kJ/day (2064 kcal/day, within the range of BMR given above), what is it in watts? (1 W = 1 J/s; 1 day = 86,400 s.)

3.4 Oxygen in a small dark cell

You are thrown into a small dark room. The door clangs shut and you fear that there are no holes or cracks for ventilation. How long will your air supply last? Reaching up, you easily touch the ceiling – seven feet high, perhaps? The floor is about six by seven, so what is the volume, allowing for your own volume? And in litres? There is no point in accuracy here, but a two-metre cube should be nearly right, i.e. 8000 l. About one-fifth of that should be oxygen, i.e. 1600 l. You recall that basal oxygen consumption is typically about 0.25 l/min, or 15 l/h, so perhaps a more convenient 16 l/h would do for your calculation if you keep still. Never mind corrections for temperature and pressure!

3.4.1 Assuming (for the calculation only!) a steady rate of consumption, how long would your oxygen last?

3.4.2 Another worrying thought: assuming that you release carbon dioxide at the same rate as you use up oxygen, what would be the percentage carbon dioxide content of the air when half that time had elapsed?

3.4.3 What is that answer when expressed as P_{CO_2} in mmHg? (Assume a barometric pressure of 760 mmHg.)

....

Naturally enough in the circumstances, a few complicating factors have been ignored. You would die before all the oxygen had gone. The rate of ventilation would rise with rising P_{CO_2} and falling P_{O_2}, and with it the work of breathing – hence also the total oxygen consumption which was assumed to be constant. Your body would accumulate excess carbon dioxide (becoming very acidotic) and not release all of it to the room.

3.5 Energy costs of walking, and of being a student

First a calculation of possible interest to slimmers, though formulated in terms of a non-obese man. The point is to express the energetic cost of walking in terms of a familiar fuel. Glucose has been chosen, although skeletal muscle actually takes up free fatty acids, acetoacetate and β-hydroxybutyrate in preference to glucose.

A 70-kg man sitting quietly expends, say, 90 kcal/h (377 kJ/h, 2160 kcal/day, 9048 kJ/day). His energy consumption walking on level tarmac at 4.5 km/h (2.8 mph) would be about 240 kcal/h (1000 kJ/h). Glucose supplies 3.7 kcal/g (15.6 kJ/g) (Table 3.1).

....

3.5.1 What would be the extra cost of walking 4.5 km (2.8 miles) at that speed as compared with simply sitting and thinking about it? Express the answer (a) in kcal or kJ and hence (b) in terms of grams of glucose.

....

Since one does not ordinarily subsist on glucose alone, here is the answer re-expressed, roughly, in terms of other foods: 65 g of bread, 21 g of butter, 28 g of potato chips, 1150 g of cucumber.

The work of walking has been measured for different speeds, different terrains, etc., but calculating it from first principles is not easy, since that requires too much knowledge of what the individual muscles are doing. One empirical finding may be noted, however. This is that walking up a slope requires an extra energy expenditure of about 7 cal (30 J) per kg of body mass for each metre that one ascends vertically, this being additional to the cost of walking horizontally. Figures similar to this have been obtained for a variety of different species of mammal. Since the mechanical work of raising a kilogram through one metre is only 9.8 J per kg of body mass (Table 1.1), there is evidently some inefficiency here, and this is discussed Section 3.9.

To estimate one's daily energy expenditure, one may look up in tables the approximate energy costs of one's principal activities and add them up, with due regard to the specific dynamic action of food, body mass, sex and so on. That is not an appropriate exercise here, but we can try to relate data we have already to daily metabolic rate.

Let us take the case of a fairly typical university student expending 2610 kcal/day (10,925 kJ/day). As an undoubted simplification, to see where it gets us, we may take the student's day as being divided between two activities, quantitatively as above – walking with an energy expenditure of 240 kcal/h (1000 kJ/h) (or doing anything else of comparable energy cost) and resting or studying with an energy expenditure of 90 kcal/h (377 kJ/h). From this information we can estimate how many hours of the day (x) is spent in walking, for (working in terms of kcal):

$$x \times 240 + (24 - x) \times 90 = 2610. \tag{3.1}$$

Hence:

$$150x = 450.$$

3.5.2 How much of the day does this student spend in walking, or in activities of equivalent energy requirement?

Is this a credible result? How might you improve on this method? It would not work as it stands for such people as farmers, coal miners and forestry workers who might use up to about 4600 kcal/day (19,000 kJ/day).

3.6 Fat storage and the control of appetite

Many adults go for long periods without much change in body mass, apart from minor short-term changes related largely to fluid balance. This implies a close match between metabolic rates, reflecting daily activities, and dietary intakes (i.e. the energy content of food minus that of faeces). Discrepancies between the two are reflected predominantly in changes in fat content. According to Table 3.1, 1 g of fat is equivalent to 9.3 kcal (39.1 kJ). Consider an individual with a (non-basal) metabolic rate of 3000 kcal/day

(12,557 kJ/day) who remains exactly in energy balance. Thus, 3000 kcal is used daily and 3000 kcal is obtained daily from food. Now suppose that the metabolic rate is decreased by 93 kcal/day (389 kJ/day) to 2907 kcal/day (12,167 kJ/day), with no change in diet. This change in metabolic rate is only 3.1% over a full day and, on the basis of figures used in Section 3.5, it is roughly equivalent to resting for 38 min of the day instead of walking for the same period. On the assumption that the unused energy is all stored as fat, the daily increase in body fat can be calculated.

3.6.1 To how many grams of fat is 93 kcal (or 389 kJ) equivalent?

This amount of energy is roughly matched by 20 g of chips or 100 g of raw potatoes.

3.6.2 If the energy intake and change in metabolic rate were maintained for one year, how much fat would the body accumulate?

Of course many adults gain or lose that much fat in a year, but such changes are not the norm. This general stability of fat content argues for the existence of control systems that match food intake very closely to energy expenditure (and to some extent *vice versa*, except that daily activity tends to be determined more by other things). Even a steady gain of 3–4 kg fat per year implies long-term regulation of energy intake to within a tolerance close to 3%. Much is known of the physiological mechanisms of hunger, appetite and satiety that govern food intake on a daily basis, but not enough to explain completely the fineness of this long-term regulation

3.7 Cold drinks, hot drinks, temperature regulation

This section has to do with appreciating the quantitative effects of certain factors that influence body temperature, including metabolic heat production and the evaporation of sweat. It starts with the effects of hot and cold drinks.

The author can comfortably take tea into his mouth at a temperature of 63 °C. A hotter drink needs to be cooled by sipping. The comfort of a warm drink in cold weather, or of a cold drink in hot weather, is obvious, but the effect on body temperature requires calculation.

As noted in Section 3.1, it takes 1 kcal, by definition, to raise the temperature 1 kg of water by 1 °C and 0.8 kcal to do the same to 1 kg of the body. In the next calculation, ignore this difference in specific heats and the small contribution of the water to the final body mass.

3.7.1 A woman weighing 60 kg drinks 600 ml (0.60 kg) of water. Its temperature is 25 °C above or below her mean body temperature. By how much would the latter change?

Clearly, the warm drink would do little for hypothermia, and the cold drink would do little for a fever. However, even the small change in temperature that has been calculated is much greater than may be needed to evoke sweating or shivering by stimulation of the hypothalamus.

3.7.2 How much heat, in kcal, must be gained or lost to bring that 0.6 kg of water drunk by the woman to her original body temperature?

We may try to give this answer more meaning by placing it in the context of metabolic heat production. Let us therefore postulate that the woman's metabolic rate happens to be 100 kcal/h (2400 kcal/day; 10,000 kJ/day). This corresponds not to basal metabolic rate, but to a low level of activity.

3.7.3 To how many hours-worth of metabolism does the previous answer correspond?

The heat of hot food and drink is not utilized to perform physiological work and is not in that sense equivalent to metabolic energy. However, hot food or drink does spare metabolic energy when the body is too cold.

We have here a 60-kg body producing heat at 100 kcal/h, which, per unit mass, is 1.67 kcal/kg h. Suppose now that the woman is in a warm, humid environment so adjusted that she neither gains nor loses heat. We may then ask how rapidly her mean body temperature would rise, remembering that it takes 0.8 kcal to raise the temperature of each kilogram by 1 °C.

3.7.4 How much would her mean body temperature rise per hour?

Fortunately such a situation is rarely maintained, but it is not necessarily as serious as might at first be supposed. This is because mean body temperature is often several degrees less than the core (including hypothalamic) temperature. Commonly about two-thirds of the body is at core temperature, with the skin many degrees colder. The answer to question 3.7.4 is merely a representative figure, much affected by physical activity, that one might have in mind when thinking generally about temperature regulation.

Ordinarily the heat is dissipated at about the rate it is produced, with radiation, conduction, convection and the evaporation of water making variable contributions to heat loss. The evaporation of water may contribute rather little in the resting person (e.g. 20%), but a great deal in exercise (e.g. 70%). What are the associated rates of water loss from the skin and airways? For simplicity, let us assume now that the above woman loses *all* of that 100 kcal/h by evaporation. The evaporation of 1 g of water requires, and so dissipates from the body, 0.58 kcal (the 'latent heat of evaporation').

3.7.5 How much water must evaporate per hour to dissipate 100 kcal/h?

Something like 10% of that might be lost in the breath (Section 5.6), so the answer implies a high rate of sweating (especially if some of the sweat runs off the skin before it can evaporate). We can secrete several litres of sweat in an hour, and the calculated rate is much less than that. Nevertheless, it is clearly a much higher rate than is generally associated with moderate activity in a cool environment. This confirms that evaporation is only a minor contributor to heat loss in such circumstances.

3.8 Oxygen and glucose in blood

The oxygen content of blood is usually expressed in terms of volumes. In fully oxygenated human arterial blood it is about 200 ml/l. This may not relate easily to concentrations of other metabolites since these are usually expressed in other units. Thus glucose is often given in mg/100 ml, being about 90 mg/100 ml (900 mg/l) in capillary blood. Given that a glucose molecule requires six oxygen molecules for its complete oxidation (reaction 3.1), one might wonder how the concentrations of the two compare when both are expressed as mmol/l. Knowing that, one would also know, for example, whether there is enough oxygen in a given volume of arterial blood to oxidize completely all the glucose that is present.

3.8.1 What is the above concentration of oxygen, 200 ml/l, re-expressed as mmol/l? (1 mmol occupies 22.4 ml at standard temperature and pressure.)

3.8.2 What is the above concentration of glucose, 900 mg/l, re-expressed as mmol/l? (The molecular mass of glucose is 180.)

It should now be clear whether or not there is enough oxygen in the blood to oxidize all the glucose completely – given that 6 mmol of oxygen is needed for each 1 mmol of glucose. Erythrocytes do not themselves metabolize glucose aerobically, but they do consume it anaerobically, with the release of lactate into the plasma, at a rate of 1.5–2 mmol/h for each litre of cells.

3.8.3 For blood containing 45% erythrocytes, what is their rate of glucose consumption in mmol/l of blood per hour when each litre of cells utilizes 2 mmol glucose per hour?

3.8.4 For a total blood volume of 5 l, what then is the rate of glucose consumption by the erythrocytes in (a) mmol/day and (b) g/day?

Glucose consumption is thus substantial, but much of the lactate produced by the erythrocytes is reconverted to glucose in the liver.

3.9 Adenosine triphosphate and metabolic efficiency

The immediate source of energy for much physiological work is adenosine triphosphate (ATP), mostly present in combination with magnesium as $MgATP^{2-}$. ATP hydrolyses thus:

$$ATP + H_2O = ADP + \text{inorganic phosphate.} \qquad [3.2]$$

The free energy change for the hydrolysis of ATP to ADP (ΔG_{ATP}) is usually estimated for intracellular conditions as -10 to -13 kcal/mol of ATP (-42 to -54 kJ/mol).

Much of our ATP is generated, by reversal of reaction 3.2, in the course of oxidative metabolism of glucose (reaction 3.1). Typically, 1 mol (180 g) of glucose yields 38 mol of ATP.

3.9.1 If the aerobic metabolism of 1 mol of glucose yields 38 mol of ATP and ΔG_{ATP} is −11 kcal/mol (−46 kJ/mol), how much energy is available from 1 mol of glucose via the hydrolysis of ATP to ADP?

One mole of glucose burnt in a bomb calorimeter yields substantially more energy than this – about 670 kcal (2800 kJ) in accordance with Table 3.1. (To obtain these figures, multiply the amount of energy per gram of glucose by 180, its molecular mass, or multiply the amount per mole of oxygen by 6.) The discrepancy between this and the answer to question 3.9.1 may be expressed in terms of the 'efficiency' of energy transfer. This is simply the answer to 3.9.1 divided by the full amount of energy that is available from 1 mol of glucose on complete combustion (multiplied by 100 to obtain a percentage).

3.9.2 What is the efficiency of energy transfer from glucose to ATP?

In the light of this result, it is clear that one cannot equate, for example, the amount of mechanical work done by a muscle and the energy from the glucose it utilizes as given in Table 3.1. In any case, there are further losses of energy as heat in the contraction process, so that the overall efficiency, defined as mechanical work divided by energy input, is even lower. The overall efficiency of the human body is least when no mechanical work is being achieved, as when a weight is being supported rather than lifted. Efficiency is near maximal in the activity of pedalling a cycle ergometer, so let us look more closely at that.

It is then typically found that the rate of energy expenditure, usually determined from measurements of oxygen consumption, rises linearly with mechanical power output – not of course from zero, since metabolism continues even when the subject is not pedalling. Taking representative values, let us suppose that the metabolic rate is 1.4 kcal/min (equivalent to 5.86 kJ/min, 98 W and about 290 ml O_2/min) when the subject is not pedalling, but 14 kcal/min (58.6 kJ/min, 977 W and about 0.29 l O_2/min) when the mechanical power output is 3 kcal/min. The overall efficiency during pedalling is equal to (3 kcal/min)/(14 kcal/min) × 100 = 21.4%. Of more interest, however, is the efficiency just of the muscular exercise (including extra activity in the heart and respiratory muscles). To calculate this it is necessary to subtract the non-pedalling metabolic rate from the metabolic rate during pedalling, before relating the result to the mechanical power output of 3 kcal/min.

3.9.3 What then is the percentage efficiency?

Provided the same units are used for both metabolic rate and mechanical power output, the linear relationship between the two is thus represented in this situation by the equation:

$$\frac{\text{metabolic rate}}{\text{while pedalling}} = \frac{\text{resting}}{\text{rate}} + \frac{(\text{mechanical power output})}{0.238}. \tag{3.2}$$

Efficiency varies very much with the type of activity, but rarely attains 25%. In a mammal, unlike a fish or frog, such energy losses are not all waste, for heat is needed to maintain body temperature.

In Section 3.5 it was noted that walking up a slope requires an extra energy expenditure of about 30 J (7 cal) per kg of body mass for each metre climbed, this being additional to the cost of walking on the level. At the same time, the mechanical work done in raising a kilogram one metre is only 9.8 J. (With '9.8' referring to g, it is particularly appropriate to work now in terms of joules rather than calories.)

3.9.4 On the basis of those two figures, what is the percentage efficiency of this component of the lifting process?

This figure exceeds the ca. 25% maximum noted above, but it is not the overall efficiency and a lower value is obtained when the total work of walking is considered. However, the calculated efficiency then has no fixed value, since it decreases with the horizontal distance that is walked per metre of ascent. In other words, it increases with increasing slope. Let us consider the 70-kg man of Section 3.5 and postulate that he ascends 1 m while walking an arbitrary horizontal distance of 10 m. According to the figures given in Section 3.5 he expends 377 kJ/h when sitting quietly and 1000 kJ/h when walking on the level at 4.5 km/h (4500 m/h). Thus, horizontal walking requires $(1000 - 377) = 623$ kJ/h above the cost of resting. Over a distance of

10 m, this amounts to $(623{,}000\,J \times 10\,m/4500\,m) = 1380\,J$. Since the man has a mass of 70 kg, the additional work involved in ascending 1 m is $(70\,kg \times 30\,J/kg) = 2100\,J$, making the total work $(1380 + 2100) = 3480\,J$. The mechanical work done in raising 70 kg is $(70\,kg \times 9.8\,J/kg) = 686\,J$.

3.9.5 So what is the percentage efficiency of the ascent?

For the energetic efficiency of the heart as a pump, see Section 4.7. For the very low thermodynamic efficiency of renal excretion, see Section 6.8.

3.10 Basal metabolic rate in relation to body size

3.10.1 A particular 70-kg man has a basal metabolic rate (BMR) of 1700 kcal/day (7000 kJ/day). A large mouse weighing 30 g has a BMR of about 4.8 kcal/day (20 kJ/day). In each case, what is the BMR expressed per kilogram of body mass (i.e. kcal/kg day or kJ/kg day)?

Expressed in this way, per unit mass, the BMR is known as the *specific BMR*. The fact that specific BMR decreases predictably with increasing body mass in mammals, from shrew to blue whale, is one of the most important generalizations in comparative physiology. One sees something similar in the general bounciness of small dogs compared with big ones – and in the more sedate behaviour of elephants as compared with mice. A similar dependence of specific BMR on body mass is seen in other groups of animals too.

As long ago as 1839 the assumption was made (by Sarrus and Rameaux) that metabolic rate in mammals of different mass (M) is proportional not to M itself, but to $M^{2/3}$. (This implies that specific metabolic rates are proportional to $M^{2/3}/M = M^{-1/3}$.) Subsequently, there has been much measurement and theorizing and we will return later to the correctness or otherwise of that two-thirds exponent. First, however, let us first see why, in mammals, the exponent cannot be 1.0. (With metabolic rate proportional to $M^{1.0}$, the specific metabolic rate is constant.)

Recall now a point of solid geometry, that for cubes of side length L the volume is proportional to L^3 and the surface area is proportional to L^2. Therefore, the ratio of surface area to volume is proportional to $L^{2/3}$. This is true of any three-dimensional body of constant shape. Moreover, if the body density is constant, the surface area is proportional to $M^{2/3}$.

Imagine two mammals that differ in body mass by a factor of 1000, but which are otherwise similar in almost every respect, including shape and fur thickness. Their surface areas must differ by a factor of $1000^{2/3} = 100$. The two mammals are postulated as both having the same metabolic rate per unit body mass (in accordance with an exponent of 1.0), so that the absolute metabolic rate of the larger animal is one thousand times that of the smaller. Assume, for ease of calculation, that the metabolic heat is lost to the environment at a rate that is proportional both to the surface area and to the difference in temperature between the body and the environment. (Heat loss from the body is not really that simple, especially if much occurs by evaporation.)

3.10.2 If the smaller mammal has a body temperature of 37 °C and the temperature of the environment is 17 °C, what must the body temperature of the larger mammal be for it to remain in heat balance?

If the assumptions behind this calculation are over-simple, the conclusion is clear enough! Over a wide range of body mass, metabolic rate could only be proportional to M if there were a huge difference in insulation or heat dissipation mechanisms between large and small mammals. However, it is obvious that small mammals are not characteristically covered either in roly-poly fat or in coats that are longer than their legs, and neither are large mammals more conspicuously provided with surfaces for heat loss.

A sensible alternative hypothesis is that metabolic rates are actually proportional to body surface area, or $M^{2/3}$. Such a 'surface rule', based on the theorizing of Sarrus and Rameaux and supported by data obtained by Max Rubner in 1883, was once generally accepted. However, it turned out not to be valid for mammals taken as a group (see below).

Even though not proportional to $M^{2/3}$, basal metabolic rates of mammals are nevertheless predictable from the following approximate equation:

$$BMR = a\,M^b, \tag{3.3}$$

where M is again body mass and a and b are constants. In logarithmic form, the equation becomes:

$$\log BMR = \log a + b \log M. \tag{3.4}$$

The two equations do not represent an exact rule, for both individual animals and individual species may diverge from them. Nevertheless, the average data for different species fall neatly around a straight line when BMR and M are plotted on a graph as logarithms. It is important to realize that the equations are regression equations that have been fitted to such data; they are descriptive and do not necessarily have a clear theoretical basis. Much has been written on their validity and interpretation.

For a range of mammals as diverse in size as shrews and elephants, the value of b has been found to be close to 3/4 rather than to the 2/3 of the 'surface rule'. The corresponding proportionality coefficient, 'a', depends on the units used, being about 70 for kcal/day and 293 for kJ/day.

3.10.3 For comparison with data in 3.10.1, what is the BMR predicted from equation 3.3 for a 70-kg man? ($70^{0.75} = 24.2$.)

3.10.4 A typical BMR for a male child of 10 kg is 600 kcal/day (2500 kJ/day). For comparison, what value is predicted from equation 3.3? ($10^{0.75} = 5.6$.)

What one might regard as reasonable agreement when making a broad survey of metabolic rates in different mammals will evidently not do in the context of predicting exact rates in adults and children. There are age-dependent factors other than body mass.

3.10.5 For use later (question 3.13.2), what is the specific BMR predicted for a blue whale of 100,000 kg? ($100,000^{0.75} = 5623$)

Comparison of the answers to questions 3.10.5 and 3.10.1 suggests one reason why blue whales can remain submerged much longer than we (or even dolphins) can.

Human metabolic rates are sometimes expressed per unit area of body surface, in accordance with the surface rule. The surface area is usually estimated from a more complicated formula than one utilizing simply $M^{2/3}$ and it is not clear that the procedure has more merit than utilizing equation 3.3 with b taken as 0.75 – or else using empirical values for a and b determined just for human beings.

For mammals as a group, many other physiological variables scale with body mass in accordance with equations of the same form as equations 3.3

and 3.4. These are known as allometric equations. Examples of such variables include skeletal mass (equation 1.9), cardiac output and glomerular
filtration rate. Some other variables, such as blood volume, heart size and
tidal volume, are proportional to body mass ($b = 1$), while ion concentrations and arterial blood pressure are nearly independent of it ($b = 0$). As a
broad generalization, some well-studied flow, rates, including respiratory
minute volume, glomerular filtration rate and renal blood flow, are proportional to BMR (i.e. $b = $ ca. 0.75), while certain frequencies (respiratory frequency, heart rate) are proportional to specific BMR (i.e. $b = -0.25$).

3.11 Drug dosage and body size

If the dosage of a drug required to produce a given effect is known for a small
laboratory animal, how does one allow for the size difference in order to estimate the dosage appropriate to a human subject? Should the dose be scaled
in proportion to body mass, or should it be related to metabolic rate? The first
might be appropriate if what matters is the maximum concentration following rapid dispersal of the drug within a particular body compartment.
Scaling in relation to metabolic rate makes sense if it is supposed that drug
clearance, by liver, kidneys or whatever, is related to general metabolism and
renal filtration. There is no exact and general answer to the question, and
specific differences in both drug response and drug metabolism may be
more important anyway. However, one can at least calculate the difference
between the two approaches.

Consider the 30-g mouse and the 70-kg man of question 3.10.1. Suppose
that a 'suitable' dosage of a drug for the mouse is found to be A units. If the
dose for the man is estimated by scaling up in proportion to body mass, then
it is $A \times 70/0.03 = 2333A$ units. The basal metabolic rates of the mouse and
the man were given as 4.8 kcal/day (20 kJ/day) and 1700 kcal/day (7000
kJ/day) respectively.

3.11.1 Scaled in proportion to BMR, what would be the appropriate dosage for a 70-kg man?

3.11.2 As a round number, what is the ratio of the two estimated doses?

An elephant was once given a dose of lysergic acid diethylamide (LSD) based
on the known effects of the drug on cats and people and scaled up on the

basis of body mass. The elephant died dramatically and unpleasantly (see Notes and Answers). Scaling of drug doses on the basis of body surface area has sometimes been advocated on the assumption that BMR is related to surface area (see Section 3.10). Because of possible interspecific differences in drug response and metabolism, any such scaling can only be regarded as a useful preliminary to empirical trials.

3.12 Further aspects of allometry – life span and the heart

> Lucretius could not credit centaurs;
> Such bicycle he deemed asynchronous.
> 'Man superannuates the horse;
> Horse pulses will not gear with ours.'
>   *William Empson* (Invitation to Juno)

An idea widely current is that our natural life span extends, roughly speaking, to a certain number of heart beats. Presumably few take it more seriously than they do our 'three score years and ten' – especially amongst those who seek to prolong their life by exercising heart and limbs! Nevertheless, we may usefully explore the matter in the context of mammals in general and thereby meet further aspects of allometry and its interpretation.

Equation 3.3 relates metabolic rate to body mass in different species of mammal. Here are two more such regression equations. The first relates the frequency of heart beat at rest to body mass (M, in kg) and the second relates life span in captivity to body mass.

$$\text{Heart rate} = 241\ M^{-0.25} \text{ beats/min.} \tag{3.5}$$
$$\begin{aligned}\text{Life span} &= 11.8\ M^{+0.20} \text{ year} \\ &= 6 \times 10^6\ M^{+0.20} \text{ min.}\end{aligned} \tag{3.6}$$

In the following calculation ignore the fact that equation 3.5 relates to resting, rather than average, heart rates.

3.12.1 According to these equations, what is the relationship between body mass and the average number of heart beats in a lifetime (heart rate × life span)?

What concerns us here is whether or not the exponent is zero, for, if it is, then the number of heart beats in a lifetime (of leisure) is about the same in large and small mammals. In fact the exponent is not zero, but it is small. Whether the discrepancy is statistically significant is another matter, for the two equations are but regression equations based on varied data. Let us consider not the statistics, but how much difference the discrepancy actually makes. This we can explore in terms of the ratio $M^{-0.05}/M^0$ ($= M^{-0.05}$). If this varies little enough with M, then 'beats/lifetime' is nearly independent of M. For a 70-kg man $M^{-0.05}$ is equal to 0.8. What about smaller mammals?

3.12.2 To take the easiest case, what is $M^{-0.05}$ in a 1-kg mammal?

The fact that $M^{-0.05}$ is not very different in the two mammals implies a similarity in average 'beats/lifetime' that would be hard to disprove. It is not to be taken too seriously. The range of body mass in mammals is of course much greater than from 1 kg to 70 kg, and $M^{-0.05}$ has more divergent values towards the extremes of M. Thus, for 0.03 kg and 100,000 kg (mouse and blue whale) the values of $M^{-0.05}$ are 1.19 and 0.56 respectively. These results illustrate the (proportionate) effects of a particular small discrepancy in the exponent b, that is to say, as between 0 and 0.05, or equally between 0.20 and 0.25 or between 0.80 and 0.85.

The next two questions are obvious ones to ask in relation to what we know about ourselves, but they are posed more in order to make a point about the allometric regression equations.

3.12.3 For a 70-kg person, what is the resting heart rate as predicted from equation 3.5? ($70^{-0.25} = 0.35$.)

3.12.4 For a 70-kg person, what is the length of life, in years, predicted from equation 3.5? ($70^{+0.20} = 2.3$.)

If either answer is unrealistic, remember that the equations represent regression lines through collections of data showing considerable scatter. This is not just due to measurement error, but reflects the considerable anatomical and physiological differences amongst species even of the same body size. Longevity happens to be better predicted if brain size, as well as body mass, is taken into account. Allometric equations summarize general trends and are not exactly predictive.

3.13 The contribution of sodium transport to metabolic rate

Sodium is continuously entering cells – in association with action potentials and sundry co-transport mechanisms for example – and it is continuously being baled out again by Na, K ATPase. Typically, the hydrolysis of one ATP molecule powers the transport of three sodium ions. Just how much of the total metabolic rate of the resting human body is normally devoted to sodium transport is hard to say, but a range of 20–45% has been suggested, with 20% as a recent estimate. It is interesting to explore the matter in the context of the considerable size dependence of metabolic rate in different species of mammal (Section 3.10). Given that cell sizes, body temperature and ionic concentrations vary little from one species to another, one might perhaps suppose that the rate of sodium pumping is about the same *per kg of body mass* in them all. At any rate, let us try out that assumption on a man and a whale.

3.13.1 Consider the man of question 3.10.1 for whom specific BMR was calculated as 24 kcal/kg day (100 kJ/kg day). Assume that 20% of the BMR is devoted to sodium transport. How much is that in kcal/kg day or kJ/kg day?

3.13.2 Consider now the blue whale of 100,000 kg for which specific BMR was calculated (question 3.10.5) as 3.9 kcal/kg day (16.5 kJ/kg day). Assuming that just as much of the specific BMR is associated with sodium transport in terms of kcal/kg day or kJ/kg day as in the man, what is that as a percentage of the total?

Could something be wrong with the assumptions? Perhaps much less energy is used for sodium transport in the resting human body than was estimated. Perhaps rates of sodium transport in different species are more nearly proportional to metabolic rate than to body mass. In the case of the kidneys, both the metabolic rate and the rate of sodium transport are closely linked to glomerular filtration rate (Section 6.8) and that scales with body mass much as does BMR.

3.14 Production of metabolic water in human and mouse

Water is produced in the catabolism of carbohydrates, proteins and fats and makes a significant contribution to body water balance. Just how much is produced in a day depends on the metabolic rate and on the proportions of

4.1.2 What percentage haematocrit corresponds to the closest possible packing of undistorted erythrocytes?

Not surprisingly, the answer is greater than any normal (human) haematocrit value. In polycythaemia, however, haematocrits may attain 70%, and this indicates that the cells cannot then be of normal shape. Obviously, erythrocytes are much more closely packed after centrifugation (with only a little plasma amongst them), so they must become considerably distorted in the process.

Perhaps the volume given for the erythrocyte (84 μm^3), or a value like it, is already familiar. If not, it is readily estimated from two other well-known quantities, as in the next calculation.

4.1.3 In men, the average red cell count in peripheral venous blood is about 5.4 million per mm³. If the corresponding (corrected) haematocrit is 44%, what is the average erythrocyte volume ('mean corpuscular volume')? (1 mm³ = 10⁹ μm^3.)

A feature of erythrocyte shape is that swelling can occur without an increase in surface area, e.g. by osmosis when the erythrocyte is placed in a dilute salt solution. Only when the corpuscle has become spherical is there any tension on the cell membrane that could lead to haemolysis. The associated increase in volume can be calculated for the erythrocyte of Table 4.1, provided that the surface area is also known. This can be estimated closely enough by treating the biconcave disc as a cylindrical disc of only half the thickness (i.e. 2.4/2 = 1.2 μm). This disc has a surface area of $(2 \times \pi r^2 + 1.2\,\mu m \times 2\pi r)$, where r is the radius (i.e. 4.15 μm). The area is accordingly 140 μm^2, which, for the erythrocyte of Table 4.1, is probably correct to within a small percentage.

Suppose now that the erythrocyte swells to become a sphere with the same surface area of 140 μm^2. Since the surface area of a sphere of diameter d is πd^2 and this area is 140 μm^2, the diameter is $\sqrt{(140/\pi)}$ = 6.7 μm. The volume, calculated as $\pi d^3/6$, is equal to $\pi \times 6.7^3/6 = 157\,\mu m^3$. The original volume, as a biconcave disc, was 84 μm^3.

4.1.4 By what factor does this erythrocyte swell in becoming spherical?

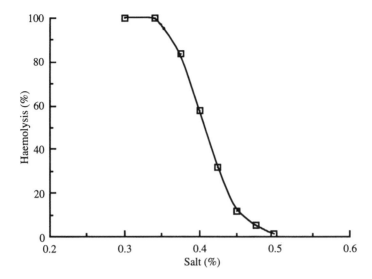

Fig. 4.2. A typical fragility curve, showing the relationship between percentage haemolysis and the concentration of the dilute NaCl solution in which the erythrocytes are suspended. Blood plasma is equivalent to 0.9% NaCl (154 mM).

This answer is relevant to the 'erythrocyte fragility test'. In this test, samples of blood are mixed with different dilutions of saline and the subsequent occurrence of osmotic haemolysis is related to salt concentration (Figure 4.2). With sufficient dilution, the erythrocytes swell to become spheres – then haemolyse as the membranes become stretched. Any given blood sample contains corpuscles of different ages, and they haemolyse at different concentrations. This is why there is a curve in Figure 4.2 and not a vertical line. Accordingly, it is useful to specify the concentration at which 50% of corpuscles haemolyse. This is typically between about 0.37% and 0.44% NaCl. Haemolysis occurs at higher concentrations when the corpuscles start more spherical, as in hereditary spherocytosis.

Let us now estimate the concentration at which the erythrocyte of question 4.1.4 would haemolyse. As an approximation, the volume of a corpuscle is inversely proportional to the salt concentration, but a more exact relationship is obtained by subtracting a constant amount from the volume. This corresponds to the volume of solids, plus a small part of the water and is found to be about 30 μm^3. Thus,

$$(\text{volume in } \mu m^3 - 30) \times \% \text{ NaCl} = \text{constant}. \tag{4.1}$$

Since the erythrocyte has a volume of 84 μm^3 in 0.9% NaCl, the constant equals $(84 - 30) \times 0.9 = 48.6$.

4.1.5 At what concentration would the erythrocyte swell to a sphere of 156 μm^3 and then haemolyse?

Is this a realistic answer? Recall that the concentration at which 50% of corpuscles haemolyse is typically about 0.37–0.44% NaCl.

In the final stage of haemopoeisis the nuclei of normoblasts become more compact (pyknotic) and are lost. How would the corpuscular volume differ if the nucleus were retained (as in reptiles, birds and other non-mammalian vertebrates)? Interphase nuclei vary in size from one kind of cell to another, but for present purposes we can choose to postulate any reasonable volume for what is just a hypothetical erythrocyte nucleus. Let us therefore start from the familiar. A small lymphocyte, as seen in a blood film, consists mostly of nucleus and has a diameter close to that of the erythrocyte. Since leucocytes tend to spread out a little on the slide, but may also shrink a little in preparation, let us take the nuclear diameter as 7 μm. (Erythrocytes and nuclei of all sorts are usually to be seen together in histological sections; are they generally of roughly similar diameter?) The volume of a sphere of diameter 7 μm is $(7^3 \times \pi/6) = 180$ μm^3. We may now imagine inserting a nucleus of this volume into the erythrocyte of Table 4.1, this having initially a volume of 84 μm^3.

4.1.6 What would be the final volume of this erythrocyte?

If this nucleated erythrocyte were spherical, it would have a diameter of 8 μm.

4.1.7 For a given haematocrit, by what factor would the haemoglobin content be reduced by the presence of the nuclei?

4.2 Optimum haematocrit – the viscosity of blood

Increasing the density of erythrocytes in blood increases its capacity to carry oxygen. On the other hand, the viscosity of the blood increases with

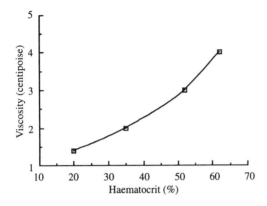

Fig. 4.3. The data of question 4.2.1 – showing how blood viscosity (in centipoises) rises with haematocrit.

increasing haematocrit and the blood flows less easily. One might guess, therefore, that there is an optimum haematocrit for oxygen carriage.

Viscosity does not increase linearly with haematocrit, but rises more and more steeply as haematocrit increases. Thus, for haematocrits of 20, 35, 52 and 62%, viscosities were found, in a particular set of experiments, to be 1.4, 2.0, 3.0 and 4.0 centipoise respectively (Figure 4.3). It does not matter here if the units are unfamiliar, but the centipoise is defined in a Note to Chapter 1. Since flow rates vary inversely with viscosities, the relative rates at which erythrocytes (and their contained oxygen) flow, e.g. through a blood vessel, may be found by dividing the four haematocrit values by the respective viscosities.

4.2.1 What are these ratios for the four haematocrit values (20, 35, 52 and 62%)?

It would seem from these calculations that the optimum concentration of erythrocytes in the blood corresponds to the normal range of values (as given in Section 4.1) – a pleasing result! This seems to be true of some other animals too (see Notes and Answers). It can be an interesting exercise to establish whether particular aspects of physiology and anatomy have been optimized during evolution (as one expects to be generally true), but the mathematics is usually far more complicated than in this example.

As so often in physiology, even this example is not as simple as might appear. A given sample of blood does not actually have a fixed viscosity that is independent of how and where it is flowing, and measurements of blood viscosity therefore depend on the kind of viscometer that is used. For example, if viscosity is estimated by timing the flow of blood through a capillary tube, then the estimate depends on the tube's internal diameter. Fortunately, the method of measurement makes less difference to the estimated optimum haematocrit; the haematocrit at which the ratio of haematocrit to viscosity is maximum works out typically as about 40–50%. Nevertheless, one should be cautious in concluding that the typical human haematocrit is exactly optimal, especially if one assumes that there is an advantage to the increased haematocrit found in people living at high altitudes. Some athletes attempt to boost their performance by increasing their haematocrits, for it is well known that a small increase (e.g. 3–6%) can enhance aerobic endurance. Too large an increase is deleterious, however.

No account was taken of the fact that oxygen is carried by the blood in simple solution as well as in combination with haemoglobin. Strictly, the above calculations should have taken account of the dissolved oxygen. However, of the typical 200 ml oxygen/l of arterial blood, only 3 ml/l is in simple solution and the correction is not worth making.

4.3 Peripheral resistance

The rate at which blood flows through any part of the cardiovascular system increases with the pressure difference between the two ends and decreases with the resistance to flow. Thus:

$$\text{flow rate} = \text{pressure difference}/\text{resistance}. \tag{4.2}$$

This commonly stated relationship neglects the kinetic energy of the blood and also its gravitational potential energy, but suffices for the present purpose. In this Section the equation is applied to the whole systemic circulation and the whole pulmonary circulation. Blood pressures are assumed to be measured at heart level, so that differences in gravitational potential energy do not need to be considered. Equation 4.2 can be rewritten as:

$$\text{cardiac output} = \frac{\text{mean aortic pressure} - \text{mean venous pressure}}{\text{resistance}}. \tag{4.3}$$

In the case of the systemic circulation it is customary to regard the central venous pressure as negligibly small compared with the arterial pressure so that, by rearrangement of equation 4.3:

$$\text{mean aortic pressure} = \text{cardiac output} \times \text{peripheral resistance.} \quad (4.4)$$

This important relationship highlights cardiac output and peripheral resistance as determinants of blood pressure. Less often given in elementary accounts are any actual values for the peripheral resistance. This is because they are not particularly memorable in themselves and have unfamiliar units (e.g. mmHg min/l). Peripheral resistance is in any case immediately calculable from the other, better-known, quantities. Thus, if the mean aortic blood pressure at rest is taken as 100 mmHg and the cardiac output is 5 l/min, then the peripheral resistance is:

$$\frac{100 \text{ mmHg}}{5 \text{ l/min}} = 20 \text{ mmHg min/l.}$$

The resistance is sometimes given in 'R units', equivalent to mmHg s/ml. Since there are 60 s per minute and 1000 ml per litre, the above value in R units is $20 \times 60/1000 = 1.2$. With a cardiac output of 6 l/min, the resistance is a convenient 1.0 R unit.

The above values of peripheral resistance become more obviously meaningful when they are compared with other values, such as those that apply in exercise and in the pulmonary circulation. If we deal with relative resistances as before, we may ignore the units.

....

4.3.1 If the cardiac output rises threefold during exercise, with mean aortic blood pressure rising from 100 mmHg to 108 mmHg, by what factor does the peripheral resistance fall? (Is the change in pressure worth including in the calculation?)

....

For the pulmonary circuit, the venous pressure is too close to the arterial pressure to be neglected. It is therefore necessary to use equation 4.3.

....

4.3.2 Taking the mean pressures in the pulmonary arteries and veins at rest as 12 mmHg and 5 mmHg respectively, what is the pulmonary resistance as a percentage of the systemic peripheral resistance given above?

....

In exercise, the pulmonary arterial pressure may rise, stay constant, or even fall slightly, depending on the nature and timing of the exercise. Changes in pulmonary vascular resistance are also variable.

4.4 Blood flow and gas exchange

For a given cardiac output, the average velocity of blood in a particular category of blood vessel – aorta, arteriole, etc. – is inversely proportional to the total cross-sectional area of the vessels in that category. The cross-sectional area of the aorta is about 4.5 cm² and the total cross-sectional area of the capillaries (variable and impossible to measure directly) has been estimated as 4500 cm².

4.4.1 Accordingly, what is the average velocity of the blood, in cm/s, in (a) the aorta and (b) the capillaries, when the cardiac output is 90 cm³/s (5.4 l/min)?

4.4.2 If a typical capillary length is taken as 0.5 mm, how long does it typically take an erythrocyte to pass through? (Ignore axial streaming, i.e. the tendency of corpuscles to travel down the middle of a vessel at a higher velocity than plasma near the endothelium; in capillaries there is often no plasma between erythrocyte and endothelium.)

That answer is for a non-exercising person; in exercise the time spent by each erythrocyte in a capillary is very much shorter. It is also shorter in smaller mammals because of their higher cardiac outputs and, as noted in Chapter 2, the whole circulation time in a shrew at rest may be a mere 3 s. That the blood velocity within capillaries is at least close to being a limiting factor in the processes of gas exchange, even in human capillaries, is suggested by the following facts. The haemoglobin of smaller mammals releases oxygen more readily at a given pH and the Bohr effect is bigger. Carbonic anhydrase activity is also greater in the blood of small mammals compared with large ones, allowing more rapid exchange of carbon dioxide.

Cardiac output may be estimated, using the Fick principle, from the measured rate of oxygen consumption and the oxygen contents of the arterial and mixed-venous blood (Section 1.3 and Chapter 2). Thus:

$$\text{cardiac output} = \frac{O_2 \text{ consumption}}{\text{arterial } O_2 - \text{mixed-venous } O_2}. \qquad (4.5)$$

The formula may be rearranged to calculate the arterio-venous difference in oxygen content from cardiac output and rate of oxygen consumption.

In Chapter 2 the mutual compatibility of certain representative textbook values for the four variables was checked. These were: cardiac output 5 l/min; oxygen consumption 250 ml/min; oxygen content of arterial blood 200 ml/l; oxygen content of mixed-venous blood 150 ml/l. Let us now assume that these figures apply to a particular individual at rest and consider the changes in fairly strenuous exercise. If cardiac output were to increase in exact proportion to oxygen consumption, then the difference between the concentrations of oxygen in arterial and mixed-venous blood would not alter. In fact the delivery of oxygen to the tissues in exercise is accomplished partly by an increase in cardiac output and partly by an increase in the amount of oxygen extracted from the blood. Let us now assume that oxygen consumption increases from 250 to 3000 ml/min, that the oxygen content of the arterial blood stays at 200 ml/l (as may happen even when exercise leads to a rise in haematocrit), and that the cardiac output increases, not proportionately from 5 to 60 l/min, but only to 20 l/min.

4.4.3 What is now the oxygen content of the mixed-venous blood?

The decrease would be greater in blood flowing through exercising muscle, and correspondingly less elsewhere.

4.5 Arteriolar smooth muscle – the law of Laplace

The law of Laplace, as applied to a cylindrical tube, relates the circumferential wall tension per unit tube length (T) to the difference between the internal and external pressures (P) and the radius (r):

$$T = Pr. \tag{4.6}$$

Here we take the external pressure as zero, so that P is simply the internal pressure, i.e. the local blood pressure in the case of a blood vessel.

This formula is often quoted in order to explain why a capillary with its thin and flimsy wall does not burst; if a capillary has a radius of 4 μm and an artery has a radius of 4 mm (4000 μm), the circumferential wall tension per unit length for a given blood pressure, like the ratio T/P, is one thousand times less in the capillary than in the artery. Less often are actual tensions calculated and units therefore specified; as expressed above, without any added numerical factor, the units are SI, namely N/m, N/m² and m for T, P and r respectively. To check that these are compatible with each other (as advocated in Section 1.3), one may rewrite equation 4.6 in terms of units:

$$N/m = N/m^2 \times m.$$

Quantification is generally only useful when there is some kind of comparison to be made and here we relate the circumferential tension in the wall of an arteriole due to blood pressure and the opposing tension that can be generated by smooth muscle (or, more specifically, the smooth muscle that is arranged in near-circular manner). In doing this we actually deal here not with tensions as such, but with mean tensions per unit area of wall thickness, S (i.e. stress). For a wall thickness w, T is wS, so that equation becomes:

$$P = wS/r. \tag{4.7}$$

The units in this equation are SI (i.e. N and m), but it may also be expressed in terms of mmHg for blood pressure and kg-force/cm² for S, both being more usual in this context. Then:

$$P \text{ (in mmHg)} = 736 \times w/r \times S \text{ (in kg-force/cm}^2). \tag{4.8}$$

Any length units can be used for wall thickness, w, and radius, r, so long as they are the same for both, since they occur in the equation only as their ratio. The constant 736, with units mmHg cm²/kg-force, is obtained from conversion factors in Table 1.1 as:

$$9.807 \text{ N/kg-force} \times 10^4 \text{ cm}^2/\text{m}^2 \div 133.3 \text{ N/(m}^2 \text{ mmHg)}.$$

We may now consider a small arteriole in which the smooth muscle layer is only one cell thick and estimate the maximum internal pressure against which circumferentially arranged smooth muscle cells can contract. For simplicity we may, for the moment, think of the arteriole wall as being made up only of such muscle cells, disregarding both the thickness of the endothelium and connective tissue and the elasticity of the elastin. Let us take a radius, r, of 15 μm and a wall (muscle cell) thickness, w, of 5 μm, and assume that the muscle cells can develop a tension of 3 kg-force/cm². Of course, all three of these vary during vasodilatation and vasoconstriction as well as from vessel to vessel. In particular, the length-tension curve for vascular smooth muscle shows a maximum tension (of perhaps 3.5 kg-force/cm²), with decreasing tensions at lengths that are longer and shorter than optimum.

4.5.1 On the basis of these figures what is the pressure, P, within the arteriole that just balances the tension in the active smooth muscle?

Actual blood pressures in the smallest arterioles vary, but are typically about 20–50 mmHg at heart level. Even allowing for an extra 100 mmHg in the feet of a standing person, it seems that the single layer of smooth muscle cells should be able to cope! This conclusion is clearly not very sensitive to the initial assumptions and simplification, but what if the vessel were much more dilated? Dilatation involves an increase in r and decrease in w, so that w/r is reduced.

4.5.2 By what factor could the ratio of wall thickness to vessel radius, w/r, decrease before contraction becomes impossible against an internal pressure of 150 mmHg, if the muscle cells still produce a tension of 3 kg-force/cm².

4.6 Extending William Harvey's argument: 'what goes in must come out'

William Harvey calculated that the heart pumps out so much blood in half an hour that only circulation makes sense (see Preface to the First Edition). We have, unlike fishes, a 'double circulation', with separate pulmonary and systemic circuits. In the long term, the flow through the two circuits must be equal.

4.6.1 A man has a blood volume of 5.00 l. Momentarily, the output from his right ventricle is 5.00 l/min and his left ventricular output is 4.95 l/min. If this small (1%) discrepancy could persist, and the blood volume were to stay constant, how much extra blood would the pulmonary vessels contain after 20 min?

The answer may be compared with the half litre of blood that would normally be present in his pulmonary blood vessels. Though the situation is unreal and the conclusion perhaps obvious, this calculation does at least highlight the need for some control mechanism to keep in long-term balance the output from each side of the heart. This is the Frank–Starling mechanism whereby the stroke volume of each ventricle increases with diastolic filling, and hence with venous return (Starling's 'law of the heart').

The same kind of argument may be applied wherever a balance of input and output must be maintained: body water balance, the balance of carbon dioxide production and loss, heat balance, the matching of transmembrane transport processes at the two sides of an epithelial cell (as in Figure 6.4), and so on.

4.7 The work of the heart

With the body at rest the oxygen consumption of the heart is commonly said to be about 8–10 ml/min per 100 g. For a heart of 300 g it would thus be about 24–30 ml/min. (This may be compared with the basal oxygen consumption of the whole body, which is commonly near 250 ml/min.) With the myocardium using as fuel about 40% carbohydrate and 60% fatty acids, 1 ml of oxygen is equivalent to about 4.8 cal (20 J). An oxygen consumption of 24–30 ml/min is thus equivalent to 115–144 cal/min (480–600 J/min). These last figures may be compared with estimates of the rate at which the heart does external work (its power output). For each side of the heart this is calculable from the following equation, in which IBP stands for the increase in mean dynamic pressure as the blood passes through the heart:

$$\text{power output} = \text{cardiac output} \times \text{IBP.} \tag{4.9}$$

The dynamic pressure of flowing blood is not identical to the blood pressure as ordinarily considered, namely the 'lateral' or 'static' pressure measured by a sphygmomanometer. Rather, it includes another component, relating to the kinetic energy due to the blood's motion. For the moment let us ignore that component in order to obtain preliminary estimates of cardiac power output and efficiency in a resting person. Accordingly, we just need to consider the increase in mean blood pressure produced by the heart, that is to say the difference between the mean arterial pressure and mean pulmonary venous pressure. When the venous pressure is close to zero, as it usually is, equation 4.9 may be simplified further to:

$$\text{power output} = \text{cardiac output} \times \text{mean arterial pressure.} \tag{4.10}$$

Following the procedures of Chapter 1, this can be analysed in terms of SI units thus:

$$W = \frac{J}{s} = \frac{N\,m}{s} = \frac{m^3}{s} \times \frac{N}{m^2}.$$

For practical purposes, rates are better expressed in terms of l/min and pressures in terms of mmHg. Equation 4.10 may be adapted accordingly, using conversion factors given in Table 1.1:

$$\frac{\text{power output}}{\text{(J/min)}} = \frac{\text{cardiac output}}{\text{(l/min)}} \times \frac{\text{mean arterial}}{\text{pressure (mmHg)}} \times 0.133$$

and

$$\frac{\text{power output}}{\text{(cal/min)}} = \frac{\text{cardiac output}}{\text{(l/min)}} \times \frac{\text{mean arterial}}{\text{pressure (mmHg)}} \times 0.032.$$

4.7.1 On the basis of these relationships, how much external work does the left heart do per minute (in J/min or cal/min) when the cardiac output is 5 l/min and the mean aortic blood pressure is 100 mmHg?

As for the right heart, the volume output is the same, but the pressures are much lower. Blood enters the right atrium at a pressure not far from zero and leaves the right ventricle at a mean pulmonary arterial pressure of about 12–15 mmHg. The external power output of the right heart is thus only 12–15% of that of the left heart. The total of the two (say 18 cal/min, 73 J/min) may be compared with the figures given above for the actual energy expenditure of the heart, namely 115–144 cal/min (480–600 J/min). The disparity may be expressed in terms of the percentage 'efficiency' of pumping, calculated as:

$$\frac{\text{external work/min}}{\text{total work/min}} \times 100.$$

4.7.2 If the total external work done per minute by the two sides of the heart is 18 cal/min (73 J/min) and the actual energy expenditure is 130 cal/min (534 J/min), i.e. within the range given, what is the percentage efficiency?

The efficiency thus calculated is fairly typical for basal conditions and it is obviously quite low. It does, however, increase substantially when the heart works harder, as in exercise, but it may be lower under pathological conditions. The variable efficiency implies that there is a poor correlation between external and total work. For a comparison with the efficiency of skeletal muscle see Section 3.9.

We may now explore the kinetic contribution to blood pressure. Like the kinetic energy of a moving solid object, this increases with the square of velocity, v, being equal to $\rho v^2/2$, where ρ is blood density. Because of this squaring, the term can become much more significant when the cardiac output is increased in exercise, and also in aortic stenosis. However, our main concern here is to show that the term is normally small at resting levels of cardiac output. To this end, we may start by ignoring the complications both of pulsatile flow and the uneven flow across the aortic cross-section and simply work with an average velocity. Strictly, it is v^2 rather than v that needs to be averaged over systole and diastole, but that would involve complications beyond the scope of this book. Let us therefore utilize the square of the mean of v rather than the mean of v^2 and see where that leads.

Now for the calculation, starting with the matter of units. With ρ in kg/m³, and v in m/s, $\rho v^2/2$ has units of $(kg/m^3 \times m^2/s^2) = kg/(m\,s^2)$. However, since $1\,N = 1\,kg\,m/s^2$, $kg/(m\,s^2)$ simplifies to N/m^2. Divided by $133.3\,N/(m^2\,mmHg)$ (Table 1.1), the expression $\rho v^2/2$ gives answers in mmHg. The mean velocity of the blood in the aorta was calculated for question 4.4.1 as 20 cm/s (0.2 m/s). The density of blood is about 1050–1064 kg/m³, but may be taken here as 1000 kg/m³.

......

4.7.3 What is $\rho v^2/2$ in mmHg when v is 0.2 m/s?

......

The point is made; the kinetic energy term is indeed very low compared with the mean blood pressure of about 100 mmHg. However, because the blood flow is actually pulsatile, the calculation is valid only insofar as it indicates the order of magnitude of $\rho v^2/2$; the answer is actually too low. It has already been pointed out that the calculation should utilize not the mean velocity, but the mean of its square, and this is less easily determined. What follows is not an attempt to obtain an exact answer, but rather an illustration of the difference between the two approaches. It is intended less as sound physiology and more as the making of an arithmetical point.

Assume that the mean velocity is again 0.2 m/s, but that flow is not actually pulsatile, but interrupted: for one-third of the time the velocity is $3 \times 0.2 = 0.6$ m/s and for two-thirds of the time the velocity is zero.

....

4.7.4 What is the mean value of v^2?

....

For comparison, recall that for steady flow at the same mean velocity, v^2 is 0.04 m²/s². With the flow regarded as intermittent, $\rho v^2/2$ is three times bigger, but that is still small enough to be neglected for the purposes above. As to the situation in exercise, the mean velocity is higher and its square is even more so. Thus, if v increases fivefold, say, then v^2 increases by a factor of twenty five.

5 Respiration

Section 5.1 shows how to correct gas volumes for variations in temperature, pressure, etc., but it is presented more as a study of the general magnitudes of such corrections and of circumstances in which they may be neglected. Sections 5.2 to 5.5 are about oxygen and carbon dioxide in blood, cells and alveoli. From a straightforward treatment of how concentrations of dissolved gases relate to partial pressures, we move on to consider cytoplasmic carbon dioxide tensions (Section 5.3), alveolar gas tensions at altitude (Section 5.4) and why the alveolar arterial carbon dioxide tensions are so much higher in mammals than in fish (Section 5.5). The latter topic is linked to that of Section 5.6 – the loss of water in expired air. Sections 5.7 to 5.9 are about breathing and the structure and dimensions of the lungs. Sections 5.10 to 5.12 are concerned with surface tensions and fluid pressures in the lungs and pleural space, and hence with pulmonary oedema. On the mathematical front, the calculations continue to use little more than simple arithmetic, but the renewal of alveolar gas (Section 5.7) provides an example of an exponential time course (for which, see also Sections 1.4 and 6.3).

5.1 Correcting gas volumes for temperature, pressure, humidity and respiratory exchange ratio

When analysing experimental results it is necessary to allow for the effects of temperature, pressure and humidity on gas volumes. Three sets of conditions are particularly interesting:

> *ATPS*: ambient temperature and pressure, saturation with water vapour at ambient temperature (as in a spirometer or Douglas bag).
> *BTPS*: body temperature, alveolar gas pressure (taken as ambient), saturation with water vapour at body temperature.
> *STPD*: standard temperature (0 °C, 273 K) and pressure (760 mmHg), dry.

In the context of the often rough-and-ready calculations of this book where the initial data tend to be 'typical values' or round numbers, the differences amongst these are not always important, and the continual need to apply conversion factors is no 'encouragement to quantitative thinking'. Indeed, many textbook accounts of respiratory physiology give volumes without specifying the relevant physical conditions. Nevertheless, the principles do need to be understood. The formulae for making conversions and corrections are given below for reference, but the main point of introducing them is to explore the general magnitude of the conversion factors so that one is in a position to judge their importance. Nothing emerges here concerning body function.

The conversion factors are based on the following relationship, in which P, V and T are respectively the pressure, volume and absolute temperature of the gas or gas mixture, n is the number of moles of gas and R is the gas constant:

$$PV/T = nR. \tag{5.1}$$

This incorporates both Boyle's law and Charles' law. Since n may include water vapour, as well as nitrogen, oxygen, etc., it does not remain constant when gas mixtures are dried or moistened. However, if n is regarded as referring only to the other gases present, then equation 5.1 can be made to apply by subtracting the water vapour pressure, $P_{(H_2O)}$, from P. Hence, for two conditions, denoted 1 and 2:

$$(P_1 - P_{1(H_2O)}) V_1/T_1 = (P_2 - P_{2(H_2O)}) V_2/T_2. \tag{5.2}$$

Table 5.1 gives water vapour pressures corresponding to saturation at various temperatures.

Conversion between V_{BTPS} and V_{STPD}

For converting between V_{BTPS} and V_{STPD}, equation 5.2 yields the following formula, in which body temperature is taken as 37 °C and the corresponding water vapour pressure is 47 mmHg. The total pressure in the alveoli is taken as the ambient (barometric) pressure, P_B, also in mmHg:

$$\frac{V_{STPD}}{V_{BTPS}} = \frac{[P_B - 47] \times 273}{760 \times [273 + 37]}$$

$$= [P_B - 47] \times 0.001159. \tag{5.3}$$

Table 5.1. *Water vapour pressures in saturated air at various temperatures*

Temperature (°C)	mmHg	kPa
0	4.58	0.61
5	6.53	0.87
10	9.20	1.23
15	12.8	1.70
20	17.5	2.33
25	23.7	3.16
30	31.8	4.24
33	37.7	5.03
35	42.1	5.62
37	47.0	6.27
40	55.3	7.37

When the barometric pressure is 760 mmHg, the ratio is 0.826, meaning that V_{STPD} is appreciably less than V_{BTPS}. Therefore, it can be important to apply a conversion factor even for approximate calculations. To a single decimal place, 0.8 is usually right, and this may be a useful figure to bear in mind, especially when barometric pressures are unknown. However, for the full range of atmospheric pressures recorded worldwide at sea level, namely 653 mmHg (870 millibars, in the eye of a typhoon) to 813 mmHg (1084 millibars, in Siberia in 1968), the conversion factor ranges from 0.70 to 0.89.

Altitude is treated more fully in Section 5.4, but for heights up to 1000 m above sea level we can take it that atmospheric pressure decreases with height h by a factor of $(1 - 0.000113h)$. Thus, 760 mmHg at sea level corresponds to 716 mmHg at 500 m.

5.1.1 According to equation 5.3, is the ratio V_{STPD}/V_{BTPS} at this pressure of 716 mmHg still equal to 0.8, i.e. to one decimal place?

Conversion between V_{BTPS} and V_{ATPS}

The formula for the interconversion of V_{BTPS} and V_{ATPS} may also be derived from equation 5.2 and is as follows, where T_{amb} is the ambient temperature in °C and $P_{amb(H_2O)}$ the ambient water vapour pressure:

$$\frac{V_{\text{ATPS}}}{V_{\text{BTPS}}} = \frac{[P_B - 47\ \text{mmHg}][273 + T_{\text{amb}}]}{[P_B - P_{\text{amb(H}_2\text{O)}}][273 + 37]}.$$ (5.4)

So long as the body is warmer than the environment, the ratio of V_{ATPS} to V_{BTPS} is less than 1.0. The ratio increases with the ambient temperature.

....

5.1.2 If the barometric pressure, P_B, is 750 mmHg, the ambient temperature is 10 °C, and the ambient water vapour pressure is 9 mmHg, corresponding to saturation at 10 °C, what is the ratio $V_{\text{ATPS}}/V_{\text{BTPS}}$?

5.1.3 Under those conditions, the ambient temperature being deliberately rather low for most laboratories, what is the percentage error if V_{BTPS} is incorrectly taken as being the same as V_{ATPS}?

....

Clearly the correction does matter for accurate work.

Equation 5.4 includes three quantities relating to the environment, namely P_B, T_{amb} and $P_{\text{amb(H}_2\text{O)}}$. If the barometric pressure, P_B, in question 5.1.2 is altered to any value between 634 and 918 mmHg, the answer works out the same as before to within ±1%. In other words, in this context, P_B does not have to be accurately known under normal conditions near sea level. On the other hand, variations in T_{amb} do make a considerable difference. The effect of this term in equation 5.4 may be seen by comparing, say, 0 and 40 °C: the ratio of $(273 + 40)$ to $(273 + 0)$ is 1.15.

Temperature also affects the maximum value of $P_{\text{amb(H}_2\text{O)}}$, i.e. the value corresponding to saturation. What if one does not know the correct water vapour pressure, but knows only that it must be somewhere between zero and the value corresponding to saturation at 37 °C (namely 47 mmHg)? Suppose that one simply takes a mid-way value of $47/2 = 23.5$ mmHg, so that, with P_B equal to 760 mmHg, $[P_B - P_{\text{amb(H}_2\text{O)}}]$ is 736.5 mmHg. The maximum percentage error in $V_{\text{ATPS}}/V_{\text{BTPS}}$ is then equal to $(760/736.5 - 1) \times 100$ or $(1 - 713/736.5) \times 100$, these being the same. Is the error likely to be generally important?

....

5.1.4 From either expression, what is the maximum percentage error in $V_{\text{ATPS}}/V_{\text{BTPS}}$ if the ambient water vapour pressure is taken as 23.5 mmHg?

....

In conclusion, V_{ATPS} and V_{BTPS} may differ appreciably, but $V_{\text{ATPS}}/V_{\text{BTPS}}$ is little affected by either P_B or $P_{\text{amb(H}_2\text{O)}}$. Variations in ambient temperature matter much more.

As equation 5.4 implies, a given mass of air increases in volume as it is inspired, through both warming and moistening. A greater volume is expired than is inspired. The two effects are no longer apparent once both volumes are expressed in the same terms (BTPS, ATPS, or STPD). However, a small discrepancy between the inspired and expired volumes may remain, because typically the volume of carbon dioxide expired is less than the volume of oxygen extracted. The ratio of carbon dioxide expired to oxygen extracted from the air is the respiratory exchange ratio (RER). In the steady state, this is the same as the whole-body respiratory quotient (RQ), usually near 0.8. How important is this effect?

As an example, suppose that dry inspired air contains 21% oxygen, 79% nitrogen, and virtually no carbon dioxide, and that dried expired air contains 16% oxygen and 4% carbon dioxide, hence 80% nitrogen. (These figures correspond approximately to RER of $4\%/[21\% - 16\%] = 0.8$, but see below for a more exact calculation.) Given that the combined percentage of oxygen and carbon dioxide falls from $(21+0)\%$ to $(16+4)\%$, it would seem that the reduction in total volume is, as a rough estimate, about 1%. In other words, it is too small to matter when no great accuracy is required. That is the main point to be made here and the reader may be content with that. However, neither RER nor the change in volume has been calculated rigorously and both need to be looked at more carefully.

To that end, let us assume that as much nitrogen is breathed in as is breathed out (as is generally the case) and that all volumes are measured dry and at one temperature. In 100 ml of inspired air there are 21 ml of oxygen and 79 ml of nitrogen. When exhaled, this 79 ml of nitrogen constitutes 80% of the dry expired air, as specified above.

5.1.5 What is the expired volume as a percentage of the inspired 100 ml?

The nitrogen is accompanied by $(4\%)/(80\%) \times 79$ ml $= 3.95$ ml of carbon dioxide and $(16\%)/(80\%) \times 79$ ml $= 15.80$ ml of oxygen (as compared with 21 ml inspired).

5.1.6 For comparison with the approximate value of 0.8 given above, what is the actual respiratory exchange ratio?

5.2 Dissolved O₂ and CO₂ in blood plasma

The concentration of a gas in simple solution increases with the partial pressure (tension) of the gas in question (e.g. P_{CO_2}, P_{O_2}) and with the solubility coefficient (e.g. S_{CO_2}, S_{O_2}). Thus, with square brackets used to denote concentrations:

$$[\text{dissolved } CO_2] = S_{CO_2} P_{CO_2}, \tag{5.5}$$
$$[\text{dissolved } O_2] = S_{O_2} P_{O_2}. \tag{5.6}$$

For a discussion of units see Chapter 1.

The following questions are to ensure familiarity both with these relationships and with normal arterial gas tensions, all of which are used in later sections. S_{CO_2} and S_{O_2} decrease with increasing temperature, but the questions utilize values appropriate to human plasma at 37 °C.

....

5.2.1 Given a normal arterial P_{CO_2} of 40 mmHg and S_{CO_2} of 0.03 mmol/l per mmHg, what is the concentration of dissolved CO_2 in plasma in mmol/l?

....

Note how low this is compared with the concentration of HCO_3 in the plasma, which is usually about 25 mmol/l.

....

5.2.2 Given an arterial P_{O_2} of 100 mmHg (approximating to normal) and S_{O_2} of 0.0014 mmol/l per mmHg, what is the concentration of dissolved O_2 in the plasma in mmol/l?

....

The answer is about the same for whole blood as for plasma. It may be compared with the total concentration of O_2 in fully oxygenated whole blood, which is about 9 mmol/l (equivalent to 200 ml/l – see question 3.8.1).

5.3 P_{CO_2} inside cells

The P_{CO_2} is generally 40 mmHg in arterial blood and about 46 mmHg in mixed-venous blood. What might it be inside a 'typical' cell? Clearly the cellular P_{CO_2} must exceed that in the nearby blood for carbon dioxide to diffuse in the right direction. In some cells, at least, it must exceed the venous value, but is this true of all cells and the 'typical' cell? One approach to this question is to consider the gradient of carbon dioxide between blood and cells and to relate

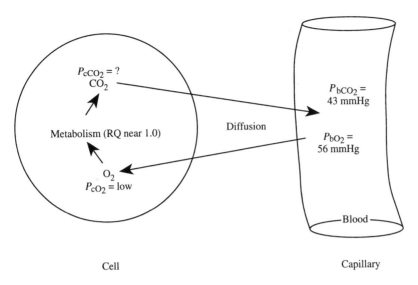

Fig. 5.1. Oxygen and carbon dioxide diffusion gradients between a cell and a nearby capillary: the gas tensions in the blood are as assumed in questions 5.3.1 and 5.3.2.

it to that of oxygen. Let us assume that the two gases move in opposite directions along the same path between cell and blood and do so only by simple diffusion (Figure 5.1). The rate of diffusion for each (along any given path) is proportional to the respective difference in partial pressure multiplied by the appropriate diffusion coefficient (for more on such coefficients, see Chapter 1). Here we do not need to know actual diffusion coefficients – only that the value for carbon dioxide that is appropriate to cells and extracellular fluid is about 20 times that for oxygen. (Note, however, that the diffusion of oxygen is facilitated somewhat by myoglobin in muscle, which lowers the ratio.)

In a steady state the rates of diffusion must also equal the rates of utilization (for oxygen) and production (for carbon dioxide). We need only to know relative, not actual, rates. The rate of carbon dioxide production divided by the rate of oxygen consumption is the respiratory quotient (RQ).

Taken together, these ideas yield the following equation, in which the subscripts b and c refer to blood and cells respectively:

$$(P_{cCO_2} - P_{bCO_2}) = (P_{bO_2} - P_{cO_2}) \times RQ/20. \qquad (5.7)$$

This may seem unhelpful at first if P_{cO_2} is also unknown, but we can at least try setting a lower limit to the latter, namely zero. For the gas tensions in blood, let us take values between arterial and mixed-venous tensions. A comparison of oxygen and carbon dioxide dissociation curves for blood in any textbook suggests that a suitable combination of P_{bCO_2} and P_{bO_2} would be about 43 and 56 mmHg, respectively. As will become apparent, the exact values matter little.

5.3.1 What is $(P_{cCO_2} - P_{bCO_2})$ when P_{cO_2} = zero, P_{bO_2} = 56 mmHg and RQ = 1?

5.3.2 Taking P_{bCO_2} as 43 mmHg, what is P_{cCO_2}?

The answer might be too high to be taken as typical since the RQ is more usually near 0.8. Moreover, P_{cO_2}, at zero, must be too low, although mitochondria can operate at oxygen tensions as low as 1 mmHg. There are estimates of the P_{cO_2} in particular tissues, for example about 2–3 mmHg in the sarcoplasm of red skeletal muscle, but the whole-body average cannot be measured. Let us take a credible, non-zero value for P_{cO_2} of 10 mmHg.

5.3.3 Accordingly, what is P_{cCO_2} when RQ = 0.8, P_{cO_2} = 10 mmHg, P_{bO_2} = 56 mmHg and P_{bCO_2} = 43 mmHg?

Compare this with the answer to question 5.3.2. One could try out various combinations of input figures in equation 5.7 without obtaining markedly different answers for P_{cCO_2}. Thus, for blood gas tensions between arterial and mixed-venous blood, and for RQ values between 0.8 and 1.0, P_{cCO_2} works out between 41 and 51 mmHg. There is no definitive figure, but it seems a fair generalization that cell P_{CO_2} must generally be higher than the nearby blood P_{CO_2} by 'a very few mmHg', or that a typical value must be close to the mixed-venous P_{CO_2}.

5.4 Gas tensions at sea level and at altitude

Inspired air, alveolar air and expired air near sea level

Let us first recapitulate a few points concerning ambient air. The atmospheric pressure at sea level varies with the weather (Section 5.1), but for present purposes may be taken as equal to the standard atmospheric pressure. This is 760 mmHg (101.3 kN/m² or 101.3 kPa – near enough 100 kPa for back-of-envelope calculations). Dry air is just under 21% oxygen, most of the rest being nitrogen with negligible carbon dioxide. The partial pressure of oxygen in dry air at standard pressure is thus 21% of 760 mmHg, i.e. 160 mmHg (21 kPa).

Inspired dry air is warmed in the airways and saturated with water vapour. The water vapour pressure at 37 °C (i.e. the partial pressure of water vapour) is 47 mmHg, so that the partial pressure of all other gases taken together is (760 − 47) = 713 mmHg. The partial pressure of oxygen, as noted also in Chapter 2, is 21% of that, i.e. 150 mmHg (20 kPa).

As already noted, the respiratory exchange ratio, RER, is the ratio of pulmonary carbon dioxide output to oxygen uptake. Equation 2.4 in Chapter 2 relates RER to the tensions of oxygen and carbon dioxide in inspired and alveolar air. After rearrangement, it becomes:

$$(\text{alveolar } P_{CO_2} - \text{inspired } P_{CO_2}) = \text{RER} \times (\text{inspired } P_{O_2} - \text{alveolar } P_{O_2}). \quad (5.8)$$

Inserting values of 150 mmHg for inspired P_{O_2} and zero for inspired P_{CO_2}, we have:

$$(\text{alveolar } P_{CO_2}) = \text{RER} \times (150 \text{ mmHg} - \text{alveolar } P_{O_2}). \quad (5.9)$$

This equation will be used later, but note again the point made in Chapter 2 that an alveolar P_{CO_2} of 40 mmHg is consistent with RER = 0.8 and alveolar P_{O_2} = 100 mmHg – all three being typical values at sea level.

The mean composition of expired air varies with the depth of breathing and is therefore less memorable than the composition of alveolar air. However, its P_{CO_2} needs to be known approximately for Section 5.5. The expired air is a mixture of alveolar air and of dead space air (which, at the start of expiration, is warmed and moistened inspired air with almost zero P_{CO_2}).

5.4.1 For a (typical) tidal volume of 500 ml and dead space of 150 ml, what is the mean P_{CO_2} in the expired air if the alveolar P_{CO_2} is 40 mmHg?

Of more practical usefulness, the calculation may be reversed – to estimate the volume of the dead space.

Altitude

Air pressure decreases with increasing altitude. Although the decrease is nearly linear up to about 1000 m (Section 5.1), the atmospheric pressure up to 7000 m can be better described as decreasing with height h by a factor of approximately $(1 - 1.16 \times 10^{-4}h + 4.4 \times 10^{-9}h^2)$. Some South Americans live at a height of about 5500 m (18,000 feet, 3.4 miles) above sea level, where the barometric pressure is about 380 mmHg. In air moistened at 37 °C, the partial pressure of water vapour is again 47 mmHg, so that the combined pressure of other gases is 333 mmHg. Of these, oxygen still constitutes about 21%.

5.4.2 What is the partial pressure of oxygen in moist inspired air at 5500 m?

The low P_{O_2} in the inspired air results in a low P_{O_2} in the alveoli also. Equation 5.9 applies again, except that the value of 150 mmHg must be replaced by the answer to question 5.4.2.

5.4.3 At the same altitude, if alveolar P_{CO_2} were to be maintained at 40 mmHg and RER were 0.9, what would alveolar P_{O_2} be?

The critical alveolar P_{O_2} at which the average individual loses consciousness on brief exposure to hypoxia is about 30 mmHg. However, the hypoxia leads to an immediate increase in pulmonary ventilation and so a fall in alveolar P_{CO_2}. P_{O_2} rises correspondingly, to a value somewhat above that just calculated. After acclimatization, the P_{CO_2} would be below 30 mmHg instead of 40 mmHg.

5.5 Why are alveolar and arterial P_{CO_2} close to 40 mmHg?

One kind of answer to this question is that P_{CO_2} is kept near that value by homeostatic mechanisms – but why was that particular value favoured during our evolution? Arterial P_{CO_2} must have been very much lower in our aquatic ancestors, as it is in present-day fish. Here we consider first why P_{CO_2} has to be low in water-breathing animals. We then look at one of the implications of having a higher arterial P_{CO_2}.

Think of a fish living in well-aerated, carbon-dioxide-free water containing oxygen at a tension of 160 mmHg as in the air we breathe. Assume for simplicity that the fish releases carbon dioxide at the same rate as it uses oxygen (i.e. RER = 1, defined as in Sections 5.1 and 5.4.). Then, as the water passes the gills, the change in its carbon dioxide content is equal and opposite to the change in oxygen content. In accordance with equations 5.5 and 5.6, each of these changes is equal to the change in partial pressure multiplied by the solubility of the gas in question (i.e. S_{O_2} or S_{CO_2}). Therefore, the partial pressures of carbon dioxide and of oxygen in the water leaving the gills are related as follows:

$$S_{CO_2} \times (P_{CO_2} - 0) = S_{O_2} \times (160 - P_{O_2}), \tag{5.10}$$

or, rearranging:

$$P_{CO_2} = S_{O_2}/S_{CO_2} \times (160 - P_{O_2}). \tag{5.11}$$

Carbon dioxide is much more soluble in water than is oxygen. Indeed, S_{O_2}/S_{CO_2} in the fish's aquatic environment would be between about $1/25$ (at $30\,^{\circ}C$) and $1/35$ (at $0\,^{\circ}C$).

....

5.5.1 What would be the maximum P_{CO_2} in the water leaving the gills, corresponding to complete extraction of oxygen?

....

The P_{CO_2} of the blood leaving the gills could, in principle, be higher than that, as a result of incomplete equilibration or through the admixture of 'venous' blood, but it would certainly have to be much lower than 40 mmHg.

Equation 5.11 may be adapted to the mammalian (air-breathing) condition by setting S_{O_2}/S_{CO_2} equal to 1.0 (because the two gases are equally 'soluble' in air, one could say) and by taking the oxygen tension of the respiratory medium (i.e. warm, moist air) as 150 mmHg instead of 160 mmHg. At this point it is useful to include a term that was taken as unity in equation 5.11 and therefore omitted, namely RER. We then have:

$$P_{CO_2} = RER \times (150 - P_{O_2}). \tag{5.12}$$

Applied to alveolar gas, equation 5.12 is the same as equation 5.9.

....

5.5.2 What is the limiting (maximum) value of alveolar P_{CO_2} for RER = 0.8, namely alveolar P_{CO_2} when alveolar P_{O_2} is zero?

....

The equation thus tells us that alveolar P_{CO_2} can be much higher in mammals than in fishes, but that there is some ill-defined upper limit set by the minimum tolerable alveolar P_{O_2}. As noted in Section 5.4, the latter exceeds 30 mmHg (corresponding to an alveolar P_{CO_2} of 96 mmHg in accordance with equation 5.12), but it is obviously less than 100 mmHg. On its own, equation 5.12 cannot explain our alveolar and arterial carbon dioxide tensions.

One reason why we have not retained the low carbon dioxide tensions of our water-breathing ancestors is probably that any surface for respiratory exchange in an air-breathing animal is necessarily also a route for evaporative water loss. Any anatomical arrangement, such as an enclosed lung, that reduces evaporation also results in higher internal tensions of carbon dioxide. The evaporation of water implies the loss of heat as well (the latent heat of evaporation, L, in kcal/mol or kJ/mol) and this is particularly important for homeotherms like ourselves. Consider the case of a mammal that is inhaling air dry and exhaling it saturated with water vapour at a partial pressure P_{EH_2O}. Then the proportion of the total metabolic rate that is devoted to making good the loss of heat through evaporation in the respiratory tract is equal to $(P_{EH_2O}L)/(P_{ECO_2}J)$, where P_{ECO_2} is the mean carbon dioxide tension in the expired air and J is the energy released in cellular respiration per mole of carbon dioxide. Air is typically expired at a temperature of about 32–33 °C, so we may take P_{EH_2O} as 37.7 mmHg (the water vapour pressure at 33 °C). In addition, we may take L as 10.4 kcal/mol of water (43.7 kJ/mol) and J as 135 kcal/mol of carbon dioxide (560 kJ/mol). (The figures for J are derived from the '4.8 kcal/l oxygen' and '20 kJ/l oxygen' of Section 3.2, with the respiratory quotient taken as 0.80.) Thus the proportion of the total metabolic rate devoted to replacing the lost heat is $(37.7 \times 10.4)/(135 \times P_{ECO_2}) = 2.9/P_{ECO_2}$.

5.5.3 On this basis, what proportion of the mammal's metabolism would be devoted to replacing the heat lost in the evaporation of water into dry inspired air if the mean P_{CO_2} of the expired air were to be (a) 2.9 mmHg and (b) more realistically, 29 mmHg?

The conclusion is clear, even though we have ignored the additional heat needed to warm the air and the work required to ventilate the lungs. It should also be remembered that P_{CO_2} would be higher in the blood than in the expired air. All of these facts strengthen the conclusion that a mammal inhaling dry air and exhaling saturated air at 33 °C would be unable to maintain P_{CO_2} as low as in most fish.

As already noted, the expired air is usually below the core body temperature. Indeed, if air is heated in the nasal passages, and water evaporates there, then the surfaces must surely be cooled somewhat, in our case typically to 32–33 °C, but lower in a very cold environment. Many mammals make use of countercurrent heat exchange in their nasal passages to cool the expired air even more than in ourselves; less water vapour is lost from these animals – and less of the heat associated with evaporation. In the short term, however, heat loss is sometimes useful. A dog mostly breathes through its well-known cold nose and exhales air at about 29 °C, but when it pants it exhales much warmer air through its mouth. Some birds reduce their arterial P_{CO_2} to a third of the normal value during panting.

5.6 Water loss in expired air

Expired air is saturated with water vapour and generally contains more of it than does inspired air. Inspired air may be far from saturated (i.e. it may have a low relative humidity), but if it is cooler than expired air its maximum water vapour pressure must be lower (Table 5.1). Thus, there is a net loss of water from the body. As already noted, the water vapour pressure at saturation in the lungs (at 37 °C) is 47 mmHg, while in expired air 33 °C it is 37.7 mmHg.

Air contracts and expands as it flows between the lungs and the environment because of changes in temperature and moisture content, which suggests that it is necessary to apply volume corrections as in Section 5.1. However, these do not materially affect the conclusions drawn from the next three calculations and may therefore be ignored.

Consider the case of someone who is breathing dry air in and moist air out at a rate of 15,000 l per day. This corresponds to a mean respiratory minute volume of 10.4 l/min, and to a fairly typical daily metabolic rate. The atmospheric pressure is, say, 760 mmHg (i.e. standard pressure). Since 1 mol (18 g) of water vapour occupies 22.4 l at standard temperature and pressure, 1 l has a mass of 18/22.4 = 0.80 g.

....

5.6.1 What is the daily rate of water loss (in g) if the air is inspired completely dry and expired saturated with water vapour at (a) 37 °C and (b) 33 °C?

5.6.2 From these two answers, how much water is saved daily, when the inspired air is dry, by breathing out air at 33 °C rather than at 37 °C?

....

Small mammals have much higher relative metabolic rates than we do (Section 3.10) and have correspondingly high rates of respiration. For desert rats it is therefore particularly worthwhile to cool the expired air.

In fact, the water vapour pressure of the environmental air is not generally zero, as in those two calculations, and it varies considerably. However, a representative figure is needed for the next calculation, and 13 mmHg is arbitrarily chosen. It corresponds to saturation near 15 °C, or under-saturation at higher temperatures.

5.6.3 What is the daily rate of water loss if the inspired air has a water vapour pressure of 13 mmHg and the expired air, saturated at 33 °C, has a water vapour pressure of 37.7 mmHg?

A typical rate of water loss in expired air is generally said to be about 400 ml/day. Actual losses depend on metabolic rate and environmental conditions.

5.7 Renewal of alveolar gas

This section makes use of normal lung volumes (tidal volume, dead space, etc.) to explore the renewal and stability of alveolar gas. At the end, the opportunity is taken to introduce an example of an exponential time course and to point out some of its properties.

Figure 5.2 is a conventional diagram illustrating, for a particular individual, lung volumes during normal breathing and with maximal inspiratory and expiratory effort. ('Lung volumes' are usually regarded as including all airways.) The tidal volume during normal breathing is shown as 500 ml, about 150 ml of which is dead space and 350 ml alveolar ventilation.

Think of this individual breathing regularly with the illustrated tidal volume of 500 ml. At the end of a normal expiration, 2500 ml of alveolar air remains in the alveoli and airways (the 'functional residual capacity'). In the subsequent inspiration this volume is mixed with 350 ml of inspired air to give a total of 2850 ml. The dilution of inspired air by alveolar air is therefore $350/2850 = 0.12$ (i.e. about one-eighth). The individual now breathes out, bringing the volume of alveolar air back to 2500 ml.

5.7.1 What proportion of the original alveolar gas would remain in the alveoli after this one breath if there were no exchanges with the blood?

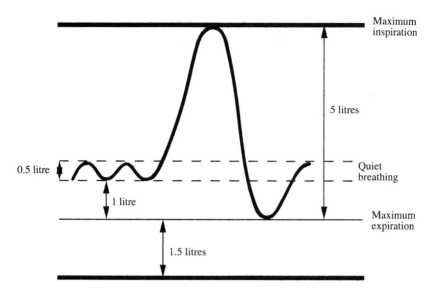

Fig. 5.2. Conventional illustration of lung volumes: the curve represents a few 'quiet' breaths followed by maximum inspiration and maximum expiration.

It is significant that so much remains because of the resulting stability in alveolar gas tensions. If all the alveolar gas were renewed in every breath, there would be large fluctuations in alveolar P_{O_2} and P_{CO_2} which would be communicated to the blood. Therefore, it is important to have a large functional residual capacity and residual volume. (It is also true that half-filled lungs need the least muscular effort in breathing.)

We now go on to consider breath-by-breath renewal of the alveolar gas, but the points to be made are more mathematical than physiological. As breathing continues, gases must in reality be exchanged between alveoli and blood, but let us ignore that fact now for mathematical simplicity, supposing that the lungs start out filled with a gas that is totally insoluble. For practical purposes helium fulfils this criterion, for its solubility in water is only about 28% of the solubility of oxygen.

For the described pattern of breathing, it has already been calculated that the proportion of alveolar gas remaining after one breath is 0.88.

5.7.2 What proportion is left unrenewed after another breath?

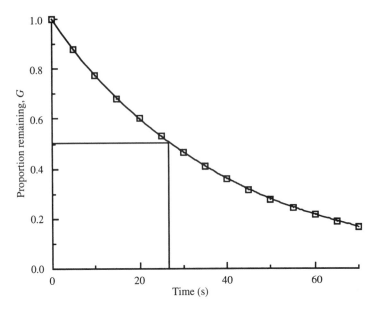

Fig. 5.3. The progressive renewal of alveolar gas over a period of 70 s at a breathing rate of 12 breaths/min. G is the proportion of the original volume remaining in the lungs. It is reduced by a factor of 0.88 at each breath. By means of the horizontal line at $G = 0.5$, and the vertical line through its intercept with the curve, the half-time, $t_{1/2}$, may be read from the time scale.

Figure 5.3 graphs the progressive changes over further breaths, with successive points given by the geometrical progression $1.00, 0.88, 0.88^2, 0.88^3, 0.88^4$, etc. The time scale has been calculated on the basis of 12 breaths/min (i.e. 1 breath every 5 s). Even though the change is actually discontinuous, a smooth curve has been drawn through the points. The curve is exponential, with an asymptote at zero. It has the following equation:

$$G = 1.00 \times e^{-kt}, \tag{5.13}$$

where G is the proportion of gas remaining, t is the time in seconds and k is the 'rate constant' (equal here to 0.0256/s, calculable as $-\ln 0.88$ divided by 5 s, the time per breath). The '1.00' is the initial value of G, analogous to Y_0 in equation 1.7.

The time required for G to fall from any value to half that value is constant and is known as the half-time ($t_{1/2}$). Here it is 27.1 s. It may be calculated as $(\ln 2)/k = 0.693/k$.

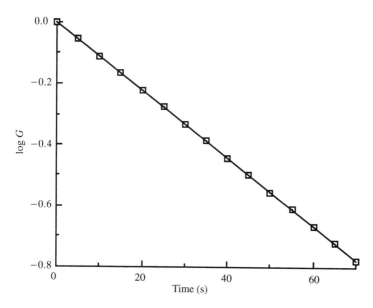

Fig. 5.4. Here the curve of Fig. 5.3 is straightened out by taking the logarithm of G, the proportion of the original alveolar gas remaining in the lungs.

If log G (i.e. $\log_{10} G$) is plotted against t, a straight line results (Figure 5.4). Its equation is:

$$\log G = - kt \times \log e = - 0.0111t. \tag{5.14}$$

The equation may also be written so as to feature the half-time instead of k:

$$\log G = - \ln 2 \times \log e \times t / t_{1/2} = - \log 2 \times t / t_{1/2}. \tag{5.15}$$

5.7.3 As a check on this equation, what is G when $t = t_{1/2}$?

Curves like those in Figures 5.3 and 5.4 may be obtained in tests of pulmonary function that involve the breathing of helium or pure oxygen. When pure oxygen is breathed, the nitrogen initially present in the lungs is gradually lost and, if the expired air is analysed, 'nitrogen washout' or 'nitrogen clearance' curves may be obtained. These are normally linear when plotted as in Figure 5.4. Departures from linearity occur when parts of the lungs are poorly ventilated.

5.8 Variations in lung dimensions during breathing

Lung volumes change during breathing (Figure 5.2), and so too do alveolar surface areas and linear dimensions. Consequently the degree of stretch of elastic fibres, the curvature of the alveolar walls and the state of the alveolar surfactant film all vary.

5.8.1 During quiet breathing as in Figure 5.2, what is the ratio of maximum to minimum lung volume?

5.8.2 In the same circumstances, if the lung spaces were to change in volume but not in shape, what would be the average variation in linear dimension, expressed as the ratio of maximum to minimum length?

Is this roughly as you imagined? The answer might be expected to apply (if only approximately) to the elastin fibres in the alveolar walls and indicates the variation in their degree of stretch. The next question concerns surface area – the area for gas exchange and the area covered by surfactant.

5.8.3 In the same circumstances, what would be the average variation in surface area expressed as the ratio of maximum to minimum?

It is the rough magnitude of change that is important here rather than the exact answer. As alveoli become inflated, their shape and appearance change in ways that defy such simple mathematical treatment.

Consider next the thickness of the surfactant-bearing film of fluid lining the alveoli. Since it is thin, its volume can be thought of as equal to the product of its surface area and its average thickness.

5.8.4 Assuming that the surface area of this film increases during inspiration by 10%, and that its volume remains constant, by what percentage does its thickness decrease?

5.9 The number of alveoli in a pair of lungs

In principle, the number of alveoli in the two lungs can be calculated as their total volume divided by the average volume of the individual alveoli. Both volumes vary during breathing and, like the actual number of alveoli presumably, they vary from individual to individual. Nevertheless, we may seek an approximate answer based on representative volumes.

For the representative individual of Figure 5.2, the total volume of gas in the pair of lungs during quiet breathing fluctuates between 2.5 and 3 l. Part of this varying volume is in the airways – a trivial 155 ml as dead space (to use the volume used in Chapter 2) and an uncertain quantity in the alveolar ducts and alveolar sacs. Looking casually at histological sections of the lung (and remembering that areas vary with the squares of linear dimensions), one may form the impression that substantially more air is in the alveoli than in the alveolar ducts and sacs from which most of the alveoli open. Accordingly, let us choose a volume of 2.5 l for total alveolar air, this being for a person of unspecified size with a corresponding total lung volume perhaps more nearly 3 l than 2.5 l.

We now need a figure for the corresponding average volume of an alveolus. The typical alveolus is roughly polyhedral, with an opening on one side, but for this approximate calculation it may be taken as spherical. Then the average volume can be estimated from the average diameter. Unfortunately, the latter is also problematical. Simple inspection of thin histological sections cannot give a clear answer, for the lung tissue may not have been fixed at a suitable degree of inflation, and most alveoli inevitably appear smaller than they really are through being sectioned eccentrically. Books and research papers give differing diameters, but for the present calculation I chose 0.25 mm. This is within the range of published averages and, being close to the diameter found by the Reverend Stephen Hales as long ago as 1731, can be chosen for that reason without favouring any one of the more recent studies. The volume of a sphere is $4\pi/3 \times (radius)^3$, or about half the cube of the diameter.

....

5.9.1 Taking a typical alveolus as a sphere of diameter 0.25 mm, what is its volume in mm³ (μl)?

5.9.2 If the volume of alveolar gas is 2.5 l (2.5×10^6 μl), how many such alveoli would there have to be?

....

Is this the order of magnitude you expected? Given the amount of uncertainty in the calculation, the answer should be regarded as correct to one significant figure.

5.10 Surface tensions in the lungs

The surface tension of the fluid lining the alveoli is important for its tendencies both to cause their collapse and to draw in fluid from the interstitial fluid and capillaries.

Laplace's formula for a sphere relates internal pressure (P) to wall tension per unit length (T) and to radius (r):

$$P = 2T/r. \tag{5.16}$$

The sphere might be, for instance, a bubble (with wall tension being surface tension) or an idealized spherical alveolus. Many books give this formula without specifying units; given as above without any numerical factor, the units are SI. Thus, if T is in N/m and r is in m, then P must be in N/m² (Pa). Surface tensions are commonly given as mN/m ($=$ dyne/cm), but a physiologist may be happier working with pressures in terms of mmHg or cmH$_2$O. In dealing with alveoli, micrometres are more convenient than metres. Here is the equation in such mixed units:

$$P \text{ (mmHg)} = 15 \; T \text{ (in mN/m or dyne/cm)}/r \text{ (in } \mu\text{m).} \tag{5.17}$$

5.10.1 Consider an air bubble in pure water with a surface tension of 70 mN/m. The radius of the bubble is 100 μm. What is the pressure, in mmHg, required to balance the surface tension?

The radius was chosen as being a round number fairly representative of human alveoli, although these vary both from one to another and with the degree of inflation. However, the fluid that lines the alveoli has a substantially lower surface tension than 70 mN/m. This is because of the presence of surfactant. The exact value of the alveolar surface tension varies during the respiratory cycle in a manner that is complicated by hysteresis. In other words, surface tension and surface area increase and decrease together, but the precise relationship between them varies with the direction of change. Generally the surface tension is between 5 and 30 mN/m. At equilibrium the surface tension is 25 mN/m. (Sometimes the surface tension in the alveoli is described as zero, but this is impossible.)

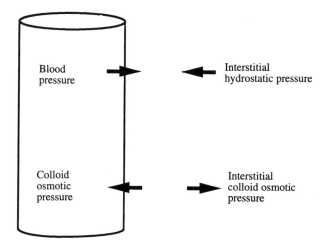

Fig. 5.5. The hydrostatic and colloid osmotic pressures influencing the flow of fluid across a capillary wall.

5.10.2 Consider another air bubble, in saline and with the surface tension lowered by lung surfactant to 25 mN/m. The radius of the bubble is again 100 μm. What is the pressure, in mmHg, balancing the surface tension?

Although an analogy between a bubble and an alveolus is implied here, an alveolus is neither perfectly spherical nor even a completely enclosed space. Moreover, it is not kept inflated by the internal pressure (which fluctuates either side of zero during breathing), but is held open by the surrounding tissues (mainly other alveoli). The pressure just calculated for the surfactant-coated bubble must therefore signify something different in relation to an alveolus. It can be thought of as a negative pressure within the thin film of fluid lining the alveolus, and one that tends both to collapse the alveolus and to draw more fluid through the alveolar epithelium from the interstitial fluid and capillaries. We consider this in the next section.

5.11 Pulmonary lymph formation and oedema

The filtration of fluid through capillary walls, whether systemic or pulmonary, depends on the balance of pressures indicated in Figure 5.5, as first proposed by E.H. Starling towards the end of the nineteenth century. The interstitial hydrostatic pressure is generally lower than the capillary blood pressure, and may even be negative. The colloid osmotic pressure is also typically lower in the interstitial fluid than in the plasma, because plasma proteins are largely retained in the capillaries during filtration of fluid through the endothelium. The picture is clearest and most familiar in relation to the systemic capillaries and in these the net pressure is usually outwards near the arterial ends, but inwards towards the venules. Within the body as a whole there is, on average, a net outwards pressure from the systemic capillaries and therefore a net outwards flow of fluid that returns to the circulation as lymph. Here our concern is with the pulmonary capillaries.

All the quantities used in this section are merely representative and some are hypothetical and hard, or impossible, to measure. In most cases the natural variability must in any case be considerable. The calculations apply to parts of the lung at heart level; in a standing person the pressure in the alveolar capillaries might be about 15 mmHg higher at the base of the lungs through the effect of gravity. Pressure gradients along the capillaries are not considered here.

In old textbooks, from the 1950s for example, one might find the colloid osmotic pressure given, correctly, as around 25 mmHg and the average blood pressure in the pulmonary capillaries, at heart level, as about 7 mmHg, but with the two other pressures of Figure 5.5 being taken as negligible.

......

5.11.1 On that basis, what is the net pressure difference across the capillary endothelium? In which direction would fluid flow?

......

It was assumed in these older accounts that pulmonary oedema would occur when alveolar capillary blood pressure exceeded colloid osmotic pressure; only when pulmonary blood pressures are abnormally high would there be any obvious role for the pulmonary lymphatic vessels. It is now apparent that the situation is more complicated and more interesting. The 'space' between the alveolar capillaries and epithelium is so small that their basement membranes are contiguous. The hydrostatic pressure there is negative, with a value estimated to be about −9 mmHg. The three pressures so far given are shown in Figure 5.6. As for the interstitial colloid osmotic pressure, it cannot be measured directly, and even the more bulky pulmonary lymph that drains from it is not easily accessible.

......

5.11.2 What would the colloid osmotic pressure of the interstitial fluid have to be for the various pressures just to balance out?

......

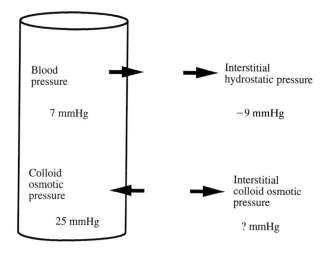

Fig. 5.6. Representative hydrostatic and colloid osmotic pressures influencing the flow of fluid across the wall of an alveolar capillary.

This is 36% of the plasma value of 25 mmHg and is in general accordance with the assumption of partial filtration of protein. For lymph to be produced, only a slightly higher value would be required and this would still be credible. In fact the concentration of protein in the pulmonary lymphatics is roughly half that of blood plasma. A round number of 10 mmHg would seem to be consistent with lymph flow and this figure is used below.

'Pulmonary oedema' is usually taken to mean the accumulation of excess fluid, not in the interstitium as elsewhere, but in the alveolar air spaces. There is normally a very thin layer of fluid lining these spaces and this bears in turn an even thinner surface film of surfactant. Because this has a surface tension, albeit substantially lower than that of a simple salt solution, there is in the film the equivalent of a negative hydrostatic pressure (Section 5.10). The actual pressure depends on the surface tension and local radius of curvature in accordance with the law of Laplace. A value of nearly 4 mmHg was calculated in 5.10.2 for the pressure required to balance the surface tension in an alveolus-like bubble of radius 100 μm, but let us here try another approach.

The air pressure in the alveoli fluctuates by about 1 mmHg either side of zero during quiet breathing and may be neglected here. In the interstitial fluid there is, using values from above, a negative hydrostatic pressure of

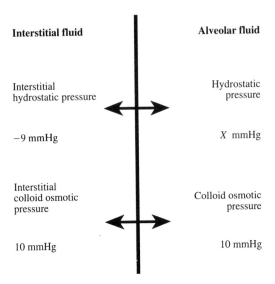

Fig. 5.7. Representative hydrostatic and colloid osmotic pressures influencing the flow of fluid between lung interstitial fluid and the fluid lining an alveolus.

−9 mmHg and a colloid osmotic pressure of 10 mmHg, both of which tend to draw water from the alveoli. Opposing these pressures is the colloid osmotic pressure of the alveolar fluid. This has not been measured, but it has been argued that it would tend to approximate that of the interstitial fluid. (The argument, in brief, is that protein moves across the alveolar epithelium by diffusion and pinocytosis and should reach about the same concentration either side.) If the volume of surface film is assumed to be steady, the hydrostatic pressure in the surface film is calculable from the balance of hydrostatic and colloid osmotic pressures. It is marked 'X' in Figure 5.7.

5.11.3 If the hydrostatic pressure in the interstitial fluid is −9 mmHg and the colloid osmotic pressures of the interstitial fluid and surface film are both 10 mmHg, what is the hydrostatic pressure in the surface film?

This negative pressure must be due to surface tension. The latter can be calculated if a value is assumed for the radius of curvature of the surface film. An obvious starting point is the radius of a typical alveolus – 120 μm perhaps. Here is equation 5.17 again, giving Laplace's formula for a sphere in convenient units:

$$P \text{ (mmHg)} = 15\ T \text{ (in mN/m or dynes/cm)} / r \text{ (in μm).} \qquad (5.17)$$

5.11.4 For the film lining an (idealized) alveolus of radius 120 μm, what is the surface tension that is equivalent to a hydrostatic pressure in the surface film of −9 mmHg?

This is close to the surface tension of pure water at body temperature (i.e. about 70 mN/m) and far too high for alveolar fluid with surfactant (for which the equilibrium value is 25 mN/m). This suggests that the calculation is based on an incorrect assumption. When an alveolus is viewed in section, the air–fluid interface is seen to be very far from smoothly spherical. Rather, its curvature varies locally, being sometimes concave and sometimes convex. The radius of curvature must be quite small in some of the nooks and crannies where the fluid tends to collect. If the surface film is to be stable and fluid is not to be drawn into the alveoli, the smallness of these radii must be offset by a lower surface tension. Surfactant is therefore essential if pulmonary oedema is to be avoided.

5.12 The pleural space

The whole pleural space, between the visceral and parietal pleurae, contains no more than about 2 ml of fluid. The elastic recoil of the lungs tends to separate the two mesothelial membranes and enlarge the space, making the pressure inside negative. Fluid therefore tends to enter through the permeable mesothelia and the lungs can only be held against the chest wall if there is an opposing mechanism for fluid removal. That must not be too effective, however, or lubrication would suffer.

We consider here the hydrostatic and colloid osmotic pressure differences (the Starling forces) between pleural space and capillary blood, doing so separately for the parietal and visceral sides. Representative pressures are summarized in Figure 5.8. Note, however, that they represent a simplification of the true situation, for the negative inside pressure not only shows a vertical gradient as in any body of fluid subject to gravity, but fluctuates with breathing; here we just take a mean pressure of −4 mmHg. As to the colloid osmotic pressure that tends to draw water into the space, this is shown as 6 mmHg. This is much lower than in plasma and corresponds to a protein concentration that is only a quarter or fifth of the plasma value.

Parietal pleurae

In the (systemic) capillaries close to the parietal pleurae, the mean blood pressure at heart level is some 22–26 mmHg. Let us take it as 24 mmHg. The

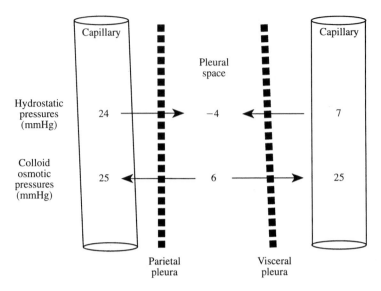

Fig. 5.8. Representative pressures influencing the flow of fluid between the pleural space and the capillaries separated from it by the parietal and visceral pleurae.

colloid osmotic pressure of the blood plasma is typically near 25 mmHg. It is assumed for present purposes that there is no significant active transport generating any further osmotic gradient across the pleural mesothelium. Pressures in the interstitial fluid surrounding the capillaries are not required for the next calculation.

5.12.1 What is the net pressure difference between systemic capillaries and pleural space, taking into account both hydrostatic and colloid osmotic pressures? In what direction does it act?

Visceral pleurae

A similar calculation may be carried out for the other side of the pleural space. The visceral pleurae are supplied with both bronchial (systemic) and pulmonary blood, but mainly the latter. The blood pressure in the pulmonary arteries is much lower than in the aorta and the pressure in the pulmonary capillaries is therefore lower than in the systemic capillaries – let us say

7 mmHg at heart level (Figure 5.8). The colloid osmotic pressure of the plasma may be taken again as 25 mmHg.

5.12.2 What is the net pressure difference between pulmonary capillaries and pleural space, taking into account both hydrostatic and colloid osmotic pressures? In what direction does it act?

One may question some of the exact pressures in Figure 5.8, but the conclusions seem clear: fluid tends to enter the pleural space on one side and leave on the other. There is thus one mechanism for keeping the pleural space small and another that ensures the presence of enough fluid for lubrication. The colloid osmotic pressure in the space must be kept low, or fluid removal would be prevented. Protein leaves the space by lymphatic drainage and pinocytosis, and some is carried out in the flow of fluid through the visceral pleurae (the process known as 'solvent drag').

6 Renal function

We start at the glomerulus – with the composition of the filtrate, the filtration rate, and how the latter is influenced by colloid osmotic pressure. Moving from glomerular to tubular processes, we apply the clearance concept to inulin, urea and other substances and then go on to consider the effects of water reabsorption on the tubular concentrations of inulin and urea. Section 6.6 quantifies the rates of filtration and reabsorption for sodium and bicarbonate in order to bring out some useful generalizations concerning kidney function.

Calculations relating to apparently 'isosmotic' reabsorption and the mechanisms and energy requirements of sodium reabsorption (Sections 6.7–6.9) place the emphasis on the bulk movements of water and solutes, rather than on the fine tuning of urine composition. Homeostasis is not neglected, however. Thus, the renal regulation of extracellular fluid volume is approached via autoregulation of glomerular filtration rate (GFR) and glomerulotubular balance (Sections 6.10 and 6.11) and there are calculations on the interrelationships amongst solute excretion, water excretion and urine concentration (Sections 6.12–6.14). In Section 6.15, certain aspects of the medullary countercurrent system are tentatively quantified in an attempt to resolve points that are often left vague in elementary accounts. Finally (Section 6.16), the abundance of mitochondria in the kidney is related to the oxygen cost of sodium reabsorption and to glomerular filtration rate.

6.1 The composition of the glomerular filtrate

Osmotic pressure

In 1844, Carl Ludwig suggested that the first stage in the formation of urine is the production of a protein-free ultrafiltrate of blood plasma, this being driven through the walls of the glomerular capillaries by the pressure of the blood. Chemical analysis of fluid from Bowman's capsule has now shown that its composition resembles such an ultrafiltrate of plasma and, in particular, plasma and capsular fluid are known to be nearly isosmotic. As long ago as 1896 G. Tammann concluded that the osmotic pressures should be similar, using essentially the following argument.

If a solution, such as blood plasma, is forced under pressure through a filter that holds back a proportion of the solutes, then the filtrate must have a lower osmotic pressure. The difference in osmotic pressure between the filtrate and the original solution would not exceed the difference in hydrostatic pressure; if it did, then filtration would cease. To complete the argument one just needs to show that the osmotic pressure of the plasma is vastly greater than the difference in hydrostatic pressures, and hence vastly greater than the difference in osmotic pressures.

The osmotic pressure of the plasma is most usually expressed in terms of the total concentration of solutes, i.e. its osmolarity or osmolality (Section 7.10). As a round number, the osmolality is 300 mosmol/kg water. The osmotic pressure at 37°C, in mmHg, can be calculated from the osmolality by means of the following relationship (see Notes and Answers for Chapter 7):

$$1 \text{ mosmol/kg water} = 19.3 \text{ mmHg.} \qquad (6.1)$$

6.1.1 What is 300 mosmol/kg water in terms of mmHg at 37°C?

This answer needs to be compared with the typical hydrostatic pressure difference between plasma and filtrate in the renal corpuscles. Putting ourselves in Tammann's position long ago, we would not know what this pressure difference is. It happens, however, that even a rough estimate will do. Let us assume that the mean aortic blood pressure at kidney level is, say, 120 mmHg and make just a small allowance for the fall in pressure along the afferent blood vessels. Then the highest likely pressure in the glomerular capillaries would seem to be about 100 mmHg, as a round number. There would be some back pressure in Bowman's space to be subtracted, but, putting aside any knowledge of its magnitude, let us take that as zero. We thus have a hydrostatic pressure difference of 100 mmHg between plasma and filtrate. This, we are assuming, represents the maximum likely difference in osmotic pressure between plasma and glomerular filtrate.

6.1.2 What is this maximum difference in osmotic pressure between plasma and glomerular filtrate, expressed as a percentage of the plasma osmotic pressure as obtained in calculation 6.1.1?

This answer should be an overestimate, not only because the back pressure in Bowman's space was neglected, but also because the net hydrostatic pressure must exceed any opposing difference in osmotic pressure in order to produce a flow of fluid. The true answer is typically about a third of what was calculated. Larger differences in the concentrations of *individual* solutes are not ruled out by this argument, but actual differences are small, except where a substance (such as calcium) is bound to plasma proteins and is therefore held back with them. The Donnan effect (Section 7.7) produces only small ionic gradients.

Note that the argument depends crucially on the acceptance of ultrafiltration as the sole process involved in the production of fluid in Bowman's space, with no contribution from active transport. The demonstration, by A.N. Richards and his colleagues in the 1920s, that fluid collected from Bowman's capsule is indeed similar to protein-free plasma was vital confirmation of glomerular filtration.

Although we have established that plasma and ultrafiltrate must have nearly the same osmotic pressure, the actual difference is nevertheless important in determining the rate of glomerular filtration. The difference is in fact the colloid osmotic pressure (oncotic pressure) of the plasma that is discussed in Sections 6.2 and 7.8.

Protein concentration

Textbooks may say, following Ludwig, that the glomerular filtrate is like the plasma from which it forms, except that it is free of protein. That the protein content is at least low is immediately evident in electron micrographs of glomeruli, where the filtrate appears clear white and the plasma (if not removed by perfusion with fixative) appears grey. Students that have been taught that, without qualification, the glomerular filtrate is protein free can be puzzled to learn that there is a mechanism in the brush borders of the proximal tubules for the recovery of filtered protein. So, we may ask, is the filtrate truly protein free? Or, if it is nearly so, why should there be a mechanism for protein reabsorption? A quick calculation answers the second question.

For this, let us postulate a low protein content for the glomerular filtrate, low enough that one might feel justified in calling the filtrate 'essentially protein free'. The protein content of plasma is about 70 g/l, so an arbitrarily postulated concentration of 0.01–0.1 g/l for the filtrate might seem to justify that phrase. (Measured concentrations do actually lie in that range, but it remains uncertain what values are typical of the healthy human kidney.)

6.1.3 A typical glomerular filtration rate is about 180 l/day (125 ml/min). How much protein would the daily 180 l of filtrate contain if the concentration in the filtrate were 0.01–0.1 g/l?

This filtered protein would be lost to the body if it were not reabsorbed. Moreover, it is not negligible as compared with the daily protein intake. The 'recommended daily allowance' of protein for a moderately active young man is about 70 g. Clearly the recovery of the calculated daily amount of protein could be life-saving during starvation.

6.2 The influence of colloid osmotic pressure on glomerular filtration rate

Glomerular filtration is driven by the hydrostatic pressure of the blood in the glomerular capillaries and it is opposed both by the back pressure within Bowman's space and by the colloid osmotic pressure (oncotic pressure) of the plasma proteins in the blood. Figure 6.1 indicates representative textbook values for each of these. (They are based on data for superficial glomeruli in rats and squirrel monkeys and it is not known how representative they are of human kidneys.)

The colloid osmotic pressure of the plasma rises along the length of each capillary as filtration of fluid concentrates the remaining protein; the value shown represents a half-way value between afferent and efferent arteriolar blood. The rise in colloid osmotic pressure can be calculated from the 'filtration fraction', itself determined from glomerular filtration rate (GFR) and renal plasma flow (RPF):

$$\text{filtration fraction} = \text{GFR/RPF.} \tag{6.2}$$

The RPF could be, say, 660 ml/min (compatible with a renal blood flow of 1.2 l/min and a haematocrit of 45%) and the GFR 125 ml/min. These figures yield a filtration fraction of 125/660 = 0.19. This means that the protein

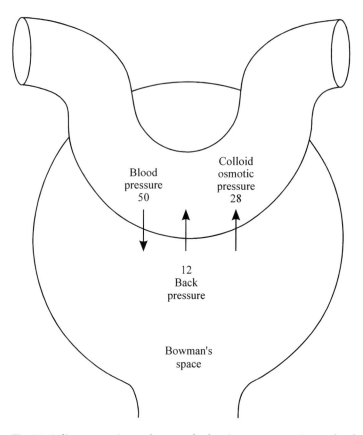

Fig. 6.1. A diagrammatic renal corpuscle showing representative textbook values for the mean hydrostatic and colloid osmotic pressures (in mmHg) that govern glomerular filtration rate.

entering the glomeruli in 1 ml of plasma leaves it in only 0.81 ml. The colloid osmotic pressure rises slightly more than the protein concentration (Section 7.8), but for present purposes we may take the two as exactly proportional to each other.

6.2.1 If the plasma entering the glomerular capillaries has a colloid osmotic pressure of 25 mmHg and the filtration fraction is 0.2, what is the colloid osmotic pressure in the plasma entering the efferent arterioles?

The colloid osmotic pressure in Figure 6.1 is midway between 25 mmHg and this value. The net filtration pressure is 10 mmHg (i.e. 50 − 12 − 28 mmHg).

For a given permeability of the filtration barrier, the GFR may be taken as proportional to the mean net filtration pressure. The next calculation shows how sensitive the GFR may be to a change in mean colloid osmotic pressure.

6.2.2 Given initial pressures as in Figure 6.1, what is the percentage increase in net filtration pressure (and hence in GFR) if the mean colloid osmotic pressure falls (by 7%) to 26 mmHg?

The point here is to demonstrate the magnifying sensitivity of GFR to colloid osmotic pressure rather than exact quantification of this change. A fuller treatment would take into account the rise in colloid osmotic pressure and fall in blood pressure along the length of each glomerular capillary, as well as a possible change in back pressure within Bowman's capsule. There is also the possibility that there are compensating changes in glomerular blood pressure – the autoregulation of GFR as discussed in Section 6.10. Dilution of the plasma proteins by an infusion into the blood of isosmotic saline does lead to a prompt diuresis in accordance with the above effect, but much of this diuresis is due to a reduction in fluid reabsorption.

6.3 Glomerular filtration rate and renal plasma flow; clearances of inulin, *para*-aminohippurate and drugs

A typical GFR for a 70-kg man is 180 l/day or 125 ml/min. GFR is measured by the 'clearance' technique, using substances that are freely filtered at the glomeruli and then neither reabsorbed nor secreted by the tubules. An ideal substance for the purpose is the plant carbohydrate inulin. Since it is not naturally present in the body, it has to be infused intravenously (first as a priming dose, then continuously to replace what is excreted). The technique makes use of the fact that, for inulin:

$$\text{rate of excretion} = \text{rate of filtration}$$
$$= \text{GFR} \times \text{concentration in plasma.} \qquad (6.3)$$

GFR is thus estimated as:

$$\frac{\text{rate of inulin excretion}}{\text{concentration of inulin in plasma}},$$

this ratio being the 'inulin clearance'. The clearance formula, as equation 1.1, is discussed in Section 1.3 with emphasis on its analysis in terms of units.

6.3.1 (Practice example) A man is infused with inulin at a rate adjusted to maintain the plasma concentration at 4 mg/l. The rate of renal excretion is found to be 0.25 mg/min (360 mg/day). What is the GFR in ml/min or l/day?

The answer is half the average normal value given above. Is the subject a small man? Has he lost one kidney? Has he a reduced number of functioning nephrons as a result of old age? GFR may be reduced to about 50% at the age of 80 and the reduction in the elderly is important in relation to drug dosage. If a drug is removed from the body only by the kidneys and GFR is low, then its concentration remains high for a longer time.

The renal plasma flow may be estimated in a similar manner, using a substance that is not only filtered at the glomeruli, like inulin, but also secreted from the blood into the tubular fluid. Such a substance is *para*-aminohippurate (PAH). If all the PAH entering the renal arteries is assumed to be excreted (as is nearly true at low concentrations), then we have the following relationship:

$$\text{renal plasma flow} \times \text{concentration of PAH in plasma}$$
$$= \text{rate of excretion of PAH}$$
$$= \text{rate of urine flow} \times \text{concentration of PAH in urine.} \quad (6.4)$$

Thus the renal plasma flow may be estimated as:

$$\frac{\text{rate of PAH excretion}}{\text{concentration of PAH in plasma}}.$$

This is the 'PAH clearance'. In fact about 10% of the blood bypasses the nephrons, so what is actually estimated this way is better described as the 'effective renal plasma flow'. A typical renal plasma flow is about 0.7 l/min (Chapter 2).

GFR and renal plasma flow are important for their effects on drug excretion. There is no place in this book for the complexities of that subject, but it is useful to consider an idealized example as a further illustration of exponential time courses. Such time courses are also discussed in Sections 1.4 and 5.7, and in their accompanying notes.

Consider the special case of a drug that is rapidly dispersed throughout the plasma and other extracellular fluid and which is lost thence only by glomerular filtration. This particular, idealized drug does not enter the cells or other body compartments. It is postulated to be neither reabsorbed nor secreted

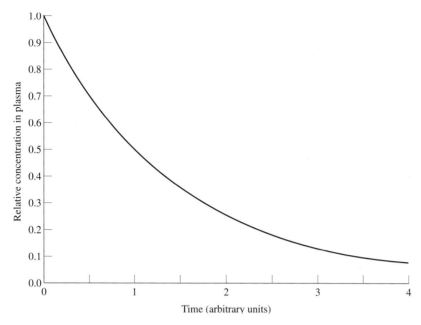

Fig. 6.2. Exponential decline in the plasma concentration of the
hypothetical drug as it is lost from the extracellular fluid by glomerular
filtration.

by the kidney tubules, and not to bind to plasma proteins. In other words, its
properties are much like those of inulin. As to its rapid dispersal and equilibration within the extracellular space, this is postulated, for mathematical
simplicity, to be much more rapid than its excretion. The excretion rate at any
moment is equal to the GFR multiplied by the concentration in the plasma
(assumed equal to the concentration in the glomerular filtrate). As excretion
proceeds, the plasma concentration falls and so too, therefore, does the rate
at which the drug is filtered. Indeed, the rate of loss from the plasma at any
instant (and hence the rate at which the concentration falls) is proportional
to the instantaneous concentration. Under such circumstances, the fall in
plasma concentration follows an exponential time course (Figure 6.2).

For a time course such as this, the period required for the concentration to
halve, i.e. the 'half-time', $t_{1/2}$, is given by the following formula:

$$t_{1/2} = (\ln 2)/k = 0.693/k, \tag{6.5}$$

where 'k' is the 'rate constant' (see also Section 5.7). For the idealized conditions described above, it is obvious that the excretion rate must be both

proportional to GFR and inversely proportional to extracellular fluid volume. Indeed:

$$k = \frac{\text{GFR}}{\text{extracellular fluid volume}}.$$

(6.6)

6.3.2 How long does it take for the drug concentration to be reduced by half if the volume of extracellular fluid is 14,000 ml and the GFR is (a) 125 ml/min, and (b) half that (62.5 ml/min)?

6.3.3 For the same extracellular fluid volume, and a GFR of 125 ml/min, how long does it take for the drug concentration to be reduced by three-quarters (i.e. half plus a quarter)?

From these answers, plus the fact that time is required for the collection of urine, it is obvious why, for the measurement of inulin clearance, it is necessary to maintain the plasma concentration by continuous infusion of inulin rather than simply giving a single injection. Excretion would be slower if the filtered drug were partially reabsorbed through the walls of the tubules and faster if the drug were actively secreted into the tubules (as is penicillin).

6.4 The concentrating of tubular fluid by reabsorption of water

By the end of the proximal convoluted tubule roughly two-thirds or more of the filtered fluid has been reabsorbed. The actual proportion depends on the individual nephron, the rate of filtration, the animal species, etc. and one can hardly claim to know a 'typical' human value. Textbook values therefore vary.

6.4.1 Suppose that inulin has been infused into a person's blood – as for the measurement of GFR (Section 6.3). In the glomerular filtrate the concentration is as in the plasma water. Inulin is neither reabsorbed from the tubular fluid nor secreted into it. By what factor is the inulin concentrated by the end of the proximal convoluted tubule if two-thirds of the fluid is reabsorbed?

Any filtered substance tends to be concentrated this way, though the tubular concentration may neverless fall because of reabsorption (as in the extreme

case of glucose), or else rise because of secretion (as with urate and penicillin).

Fluid reabsorption continues along the nephron (except that only minor
water movements occur through the walls of the ascending limb of the loop
of Henle) and only a tiny proportion of the filtered water emerges in urine.

**6.4.2 Suppose that the GFR is 125 ml/min and that urine is produced at a
rate of 1.25 ml/min. By what factor is the inulin concentrated in the urine?**

It is no small point that inulin, a substance foreign to the body, should
become concentrated in the urine like this. It is still unclear why vertebrate
kidneys filter and reabsorb so much fluid, but a clear advantage to this seemingly wasteful system is that it ensures the excretion of any substance, like
inulin, for which no special transport system has evolved.

**6.4.3 The concentration of glucose in plasma is about 5 mmol/l. The renal
reabsorption of glucose can be inhibited by phlorizin. With complete
inhibition, what would be the concentration of glucose in the urine,
assuming flow rates as given in question 6.4.2?**

This is quite high compared with the osmotic concentrations of both plasma
(300 mosmol/kg water) and urine (30–1400 mosmol/kg water). An osmotic
diuresis would result (as it does in diabetes mellitus, where glucose reabsorption is typically incomplete) and therefore the concentrating of the
glucose would actually be less than has been calculated. (For further discussion of osmotic diuresis, see Section 6.13.)

6.5 Urea: clearance and reabsorption

It has not always been clear that the first process in urine formation is filtration at the glomeruli. Rudolph Heidenhain did not believe in filtration and,
in 1883, revived instead an older theory of glomerular secretion. One reason
for his views was the very high value for the GFR that was required to account
for the known rate of urea excretion. This he calculated on the assumptions:
(1) that urea is filtered at the same concentration as in plasma; and (2) that
the rate of excretion is the same as the rate of filtration. In accordance with
these assumptions, equation 6.3 should apply just as for inulin:

rate of excretion = rate of filtration

$$= \text{GFR} \times \text{concentration in plasma.} \qquad (6.3)$$

The normal concentration of urea in plasma averages about 4.5 mmol/l (270 mg/l) with a range of about 2.5–7.5 mmol/l. The excretion rate is typically about 270–580 mmol/day.

6.5.1 On the above assumptions, what GFR (in l/day) would correspond to an excretion rate of 450 mmol/day and a plasma concentration of 4.5 mmol/l?

Using his own figures, Heidenhain obtained a roughly similar estimate for GFR (70 l/day). Although we now know that the true GFR (for a 70-kg man) is typically close to 180 l/day, Heidenhain regarded his estimate as too preposterously high to be real. His estimate was actually far too low because much of the filtered urea is reabsorbed into the blood – a fact which surprises those who do not expect reabsorption of an excretory product.

What was effectively calculated for question 6.5.1 is what is known as the 'urea clearance', though this is usually expressed in ml/min. The actual urea clearance depends on the rate of urine flow, being least when the flow of urine is slow and rising with increasing flow rates (Figure 6.3). The answer to question 6.5.1 (equivalent to almost 70 ml/min) is typical of urine flow rates exceeding 2 ml/min.

We have seen that the urea clearance provides an underestimate of GFR, this being because much of the filtered urea is reabsorbed. The proportion that is reabsorbed can be calculated from the following formula; it applies to any substance that is filtered at the same concentration as it occurs in plasma.

$$\text{Fraction reabsorbed} = 1 - (\text{clearance/GFR}). \qquad (6.7)$$

6.5.2 If the urea clearance is 70 ml/min and the GFR is 125 ml/min, what fraction of urea is being reabsorbed?

As to the mechanism of reabsorption, urea, like inulin (Section 6.4), is concentrated in the tubules by the reabsorption of water. The resulting concentration gradient between lumen and surrounding blood leads to outward diffusion through the permeable walls of the tubules. As a result, the concen-

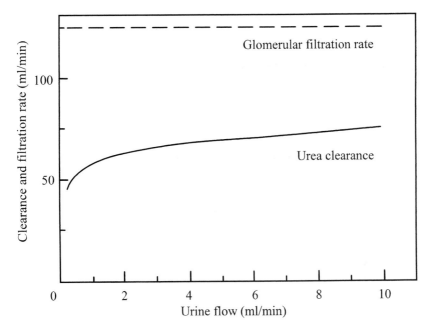

Fig. 6.3. The dependence of urea clearance on rate of urine flow. The broken line shows a constant glomerular filtration rate of 125 ml/min.

tration of urea in the urine is elevated less than it would otherwise be. Let us consider now only what happens in the proximal convoluted tubules, for beyond these the situation is complicated by the medullary gradients of urea concentration that are so important to the countercurrent mechanism. In the proximal convoluted tubules the maximum conceivable proportion of filtered urea that could be reabsorbed by the passive mechanism just indicated (i.e. the limiting value) would correspond to virtually no elevation of the urea concentration compared with plasma.

6.5.3 What is the maximum proportion of urea (i.e. the limiting value) that could be reabsorbed in the proximal tubules by this mechanism, assuming that two-thirds of the filtered water is reabsorbed there?

Compare this with the answer to question 6.5.2.

In the event of total kidney failure urea excretion ceases, but the hepatic production of urea continues. Urea therefore accumulates in blood and tissues.

6.5.4 Suppose that urea is produced within the body at a rate of 450 mmol/day following complete renal failure (so that none is excreted), and that it is distributed evenly through all compartments of body water. If the volume of this water is 45 l, by how much does the concentration of urea rise in one day (in mmol/l)?

6.6 Sodium and bicarbonate – rates of filtration and reabsorption

It is important to realize just how much sodium or sodium chloride is reabsorbed, and the first two calculations are about this. Consideration of bicarbonate reabsorption and acid excretion then leads to important generalizations about kidney function.

The rate at which any solute is filtered may be calculated as the product of the GFR and the concentration of the solute in the filtrate. The latter, for anything but protein and substances strongly bound to protein, may be taken as approximating the concentration in blood plasma. (Small differences due to the Donnan effect are discussed in Section 7.7.) For the concentration of sodium in the filtrate, a round value of 150 mmol/l will do; if this seems a little high for typical plasma, remember that the sodium there is diluted somewhat by the volume of plasma proteins (Section 7.9). As already noted, the GFR of a textbook 70-kg man averages 180 l/day.

6.6.1 According to these figures, how many moles of sodium are filtered in a day?

Expressed thus, in terms of moles, the quantity is hard to visualize, so it may help to think in terms of a mound of salt. The filtered sodium is accompanied mainly by chloride and, to a lesser extent, bicarbonate, but for the purpose of visualization we may consider it all as sodium chloride. The formula mass of NaCl is 58.5.

6.6.2 Expressed in terms of kilograms of its chloride, how much sodium is reabsorbed by the kidneys in a day?

The 27 mol of sodium filtered in a day compares with a daily excretion, on a typical western diet, of 0.08–0.2 mol. These quantities are so very different

obviously limiting is the haemoconcentration accompanying glomerular filtration, i.e. the rise in both erythrocyte concentration (haematocrit) and plasma oncotic pressure.

Of course, the kidneys do not use oxygen just in reabsorbing sodium. Their total oxygen consumption is thus somewhat more than was obtained in calculation 6.8.2, but probably only by a litre or so per day. However, even this total does not represent the whole cost of urine production, for the heart contributes too, driving blood through the kidneys and fluid through the glomerular filters. With the body at rest the oxygen consumption of a 300-g heart is about 24–30 ml/min (Section 4.7), or 35–43 l/day. About one-fifth of the systemic cardiac output goes to the kidneys.

6.8.6 If the rates of oxygen consumption of the kidneys and heart are respectively 24 l/day and 40 l/day, and if the blood flow to the kidneys is one-fifth of the cardiac output, what is the total cost of urine production in l oxygen/day?

Although two hypothetical values for the oxygen content of renal venous blood were postulated in relation to question 6.8.5 (i.e zero and 150 ml/l), the actual oxygen content has yet to be addressed. A representative value may be obtained using the Fick Principle. The formula given in Chapter 1 (equation 1.3) may be adapted to the present situation as follows:

$$\text{renal blood flow} = \frac{\text{renal oxygen consumption}}{\text{arterial } O_2 \text{ content} - \text{venous } O_2 \text{ content}}. \quad (6.8)$$

Let us take the total oxygen consumption of the kidneys as 16 ml/min, the oxygen content of the arterial blood as 200 ml/l and the renal blood flow as 1.25 l/min.

6.8.7 What is the oxygen content of the renal venous blood?

6.9 Mechanisms of renal sodium reabsorption

What can we learn about the mechanisms of renal sodium reabsorption from its cost in terms of energy or oxygen? The processes of solute transport in different parts of the nephron and collecting duct are many. Thus, sodium ions

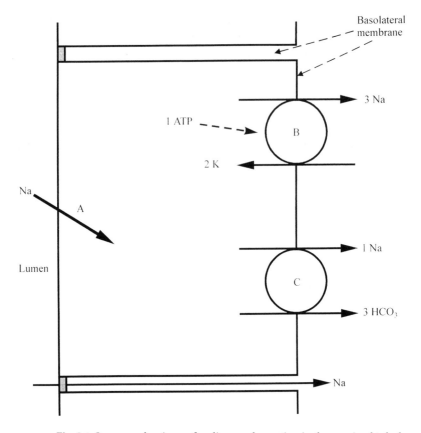

Fig. 6.4. Some mechanisms of sodium reabsorption in the proximal tubule. From the lumen, to the left, sodium enters the cell in various ways down its electrochemical gradient (A). It leaves through the basolateral membrane via the sodium pump (B) and by electrogenic cotransport with bicarbonate (C). ('Cotransport' signifies that the sodium and bicarbonate are transported together, in the same direction and on the same carrier in the cell membrane. The cotransport is called electrogenic because more bicarbonate ions are transported this way than sodium ions, so that there is a net transfer of charge.) Sodium also crosses the epithelium through the tight junctions (the paracellular pathway).

may pass through the apical (luminal) cell membranes alone through channels, or in association with other solutes on carriers (accompanying glucose or phosphate, exchanging with hydrogen ions, etc.). Having entered the cells, some of the sodium leaves through the basolateral membranes by electrogenic cotransport with bicarbonate, but most is expelled by the sodium pump (Figure 6.4). In other words, most of the sodium passing through the cells is actively transported through the basolateral membranes in exchange for potassium and at a cost of one ATP molecule for each three sodium ions transported. How does this square with the idea (Section 6.8) that about 29 sodium ions are associated with the consumption of 1 oxygen molecule?

To answer this we need to know the relationship between oxygen consumption and ATP production, but this depends on the nature of the respiratory substrate. The renal medulla utilizes mainly glucose, but in the cortex, where most of the sodium reabsorption occurs, the main substrates are fatty acids. When the substrate is a fatty acid, the ratio of ATP molecules produced per oxygen molecule depends on which fatty acid is involved, but we may take palmitate as representative. This has an ATP/O_2 ratio of 5.6 (as compared with 6.3 for glucose).

6.9.1 Assuming that 29 sodium ions are associated with the consumption of 1 oxygen molecule and that the ATP/O_2 ratio is 5.6, what is the ratio of sodium ions reabsorbed to ATP molecules synthesized and broken down?

As a round number, the answer is the same for palmitate and glucose and it is well above the ratio of 3 that typically corresponds to transport by Na,K-ATPase. This may be surprising inasmuch as it is easy to see how the ratio could be less than 3 – through back-leakage of Na and through consumption of ATP in other ways. The discrepancy may be accounted for in terms of two processes that do not directly involve Na,K-ATPase. One of these is the reabsorption of sodium between the cells, i.e. through the paracellular pathway. This is an important route for sodium reabsorption in both the proximal tubules and the loops of Henle. The other process not involving Na,K-ATPase has been mentioned already, namely the cotransport of sodium with bicarbonate through the basolateral membranes. However, the proportion of sodium reabsorbed in this way is quite small, as will be shown next.

We now calculate an upper limit for the proportion of sodium that might be reabsorbed by the cotransport of sodium and bicarbonate through the basolateral membranes of the epithelial cells. As already noted (Section 6.6), sodium and bicarbonate are reabsorbed by the whole kidney in roughly the same molar ratio as they occur in plasma. This ratio (i.e. Na/HCO_3) is about 6. Basolateral electrogenic cotransport of the two ions involves the movement of three bicarbonate ions with each sodium ion (Figure 6.4).

6.9.2 If *all* the bicarbonate reabsorbed in the kidneys were to cross the basolateral membranes of the epithelial cells of the tubules in association with sodium, what percentage of the total reabsorbed sodium would have to be reabsorbed with bicarbonate (rather than by Na,K-ATPase)?

The percentage of sodium reabsorbed this way must be even lower if some bicarbonate is reabsorbed in other ways.

6.10 Autoregulation of glomerular filtration rate; glomerulotubular balance

The rate at which urine is produced is necessarily equal to the GFR minus the net rate of reabsorption of fluid from the renal tubules. Thus, a rate of urine production of 1 ml/min might correspond to a GFR of 125 ml/min and a rate of reabsorption of 124 ml/min, i.e.,

$$1 \text{ ml urine/min} = 125 \text{ ml/min (filtered)} - 124 \text{ ml/min}$$
$$\text{(reabsorbed)}.$$

6.10.1 What is the rate of urine flow if (a) the GFR rises to 126 ml/min without change in the rate of reabsorption, and (b) the GFR stays at 125 ml/min and the rate of reabsorption falls to 123 ml/min?

Since urine flow changes so much in the stated circumstances, one would expect there to be mechanisms controlling both the GFR and the rate of fluid reabsorption. (Note that, although one may postulate rates like those above, the rate of reabsorption cannot be directly measured and the change in GFR is too small to be detected experimentally.)

Two phenomena have been revealed experimentally that correspond to the two requirements for control thus identified. One is the autoregulation of GFR, whereby GFR remains nearly constant in the face of variations in the blood pressure in the renal arteries. (Good autoregulation only occurs over a limited range of arterial blood pressure and other mechanisms can override it, so that, for example, the GFR may fall during exercise. Autoregulation of GFR is linked to autoregulation of renal blood flow.) The second phenomenon is known as glomerulotubular balance and involves adjusting the rate of fluid reabsorption to match such changes as do occur in GFR. The difference between GFR and rate of fluid reabsorption (i.e. the rate of urine flow) is thus far less variable than might otherwise be. Both autoregulation and glomerulotubular balance are sometimes described as total, but it would be surprising if that were entirely true. Indeed, the next section explores the idea that a slight 'imperfection' in both mechanisms may actually be important to homeostasis.

6.11 Renal regulation of extracellular fluid volume and blood pressure

Many mechanisms contribute to the regulation of extracellular fluid volume. These involve hormones (angiotensin, aldosterone, atrial natriuretic hormone, vasopressin), renal reflexes and the colloid osmotic pressure of the blood plasma. The mechanism that concerns us here is the direct effect of arterial blood pressure on the rate of urine production. Some people regard this as the most important long-term determinant of blood pressure, the idea being as follows. Raised blood pressure acts directly on the kidneys to increase the excretion of both salt and water from the extracellular fluid; this leads in turn to a reduction in the plasma volume and 'fullness' of the cardiovascular system, and so to restoration of blood pressure. Of course, there are many other factors and control systems to complicate the story, including those already mentioned.

Again we consider the necessary dependence of urine flow on the difference between GFR and rate of fluid reabsorption (Section 6.10). As a starting condition, suppose that the mean arterial blood pressure of an individual is 100 mmHg, that the GFR is 125 ml/min, and that urine is being formed at 1 ml/min.

6.11.1 What is the rate of fluid reabsorption?

Now suppose that the mean arterial blood pressure rises from 100 to 140 mmHg (the latter being an illustrative value that is not essential to the argument). Bearing in mind the phenomenon of autoregulation of GFR, and assuming that this cannot generally be completely effective, let us suppose that the GFR rises not in proportion to blood pressure (i.e. to $140/100 \times 125 = 175$ ml/min), but only to 130 ml/min (an arbitrary rise of only 4%). Finally, let us postulate that fluid reabsorption rises too (glomerulotubular balance), but by 3% rather than the full 4%, to 128 ml/min. The only figures needed for the next calculation are these new rates of filtration and reabsorption, and the original rate of urine flow (1 ml/min).

6.11.2 By what factor does the rate of urine production increase in response to the rise in blood pressure?

It has been known since 1843 that urine flow increases with perfusion pressure – in isolated kidneys and *in vivo*. The phenomenon is called 'pressure diuresis'. It is accompanied by increased excretion of sodium ('pressure natriuresis'). Although based on arbitrary figures, the above calculations suggest that pressure diuresis is almost inevitable, but is a small rise in GFR really responsible? Pressure diuresis is said to occur in the absence of any detectable rise in GFR, but detection of a 4% change is not easy. Be that as it may, there is undoubtedly more to the phenomenon. Thus, increased pressure in the vasa recta has been found to reduce fluid reabsorption in the papillae. Secretion of urodilatin, a potent diuretic and natriuretic hormone, in response to raised intrarenal blood pressure could also be involved.

6.12 Daily output of solute in urine

The total amount of solute excreted daily in the urine expressed in milliosmoles (the 'excreted solute load') is a valuable concept in the clinical interpretation of water imbalance. Representative values for the daily urinary solute output are required in Sections 6.13 and 6.14.

The main solutes in urine are urea and inorganic ions. Table 6.1 shows typical excretion rates for the substances most generally abundant in the urine. Actual ranges depend very much on diet and are by no means as well defined as the table implies. Thus, to take the case of sodium, intake and excretion may be as high as 500–600 mmol/day in parts of Japan and they may be as low as 2–10 mmol/day in the Amazon jungle and highland New Guinea, where plants and soil are very low in salt. The rate of urea excretion increases with the protein content of the diet, but when the latter is zero the rate of urea excretion only falls to about 100 mmol/day. This is because of the breakdown of body protein.

An estimate of the typical range of solute output may be obtained by summing all the minimum values in Table 6.1 and also all the maximum values. The results will suffice for present purposes, but the method is hardly

Table 6.1. *Typical renal excretion rates on a western diet*

Substance	Excretion rate (mmol/day)
Urea	270–580
Sodium	80–200
Chloride	80–200
Potassium	40–150
Ammonium	30–70
Phosphate	26–42
Sulphate	18–28

accurate. It could give too great a spread inasmuch as the two extreme values would require that every solute be simultaneously minimum or maximum. On the other hand, the ranges for the individual solutes are not themselves extremes. Because the units in the table are mmol/day instead of mosmol/day as required, the totals should be reduced by roughly 10% to allow for non-ideal behaviour of the urine in terms of its physical chemistry. However, a small percentage can be added back to allow for untabulated solutes. It should thus suffice to sum the figures as they are, so long as the estimates are regarded as approximations.

6.12.1 On the basis of Table 6.1, what seems to be a likely range for the solute output in mosmol/day?

Daily solute output may also be obtained directly from measurements of osmotic concentrations and volumes of pooled urine samples.

The ammonium ions and the urea of Table 6.1 both contain nitrogen and this may now be related to the dietary protein from which most of it ultimately derives. Since each molecule of urea contains two nitrogen atoms, the rate of nitrogen excretion (ignoring untabulated nitrogen compounds such as creatinine) is, according to Table 6.1, about 570–1230 mmol/day, or 8–17 g/day. The number of grams of protein containing this amount of nitrogen is readily estimated, given the fact that, for many proteins, 1 g of nitrogen is equivalent to 6.25 g of protein.

6.12.2 To how much protein does this range of daily nitrogen excretion correspond?

Daily allowances for protein in the diets of young men and women, recommended by the Department of Health, are in this range.

6.13 The flow and concentration of urine

Section 6.12 introduced the notion of daily solute output or excreted solute load. We may take values on a western diet as being generally in the range of 500–1200 mosmol/day. The daily solute output is equal to the product of daily urine volume and average solute concentration. Alternatively:

$$\text{urine concentration in mosmol/l} = \frac{\text{solute output in mosmol/day}}{\text{urine flow in l/day}}. \quad (6.9)$$

Although formulated in terms of the whole day, the equation may of course be applied to shorter time scales. (Osmotic concentrations are given elsewhere as 'mosmol/kg water' as that is usually most appropriate, but here it is more convenient to work in terms of volumes.)

So long as the rate of solute output remains constant, concentration varies inversely with flow rate. The next three questions illustrate this in the context of normal human values.

6.13.1 An individual produces isosmotic urine, so that its concentration, like that of the plasma, is about 300 mosmol/l. If the solute output is 750 mosmol/day, what is the rate of urine production?

The daily rate of urine production is said to be more typically nearer half this.

6.13.2 So, for the same solute output but half the rate of urine production, what would the (more typical) concentration be?

6.13.3 For the same solute output, what concentrations would correspond to the two fairly extreme flow rates of (a) 15 l/day and (b) 0.6 l/day?

The unusually high flow rate of 15 l/day could be associated with diabetes insipidus or the excessive water drinking that occurs in some psychiatric disorders. In the short term, however, the same rate ($\equiv$ 10 ml/min) is commonplace after a single large drink. For a complete curve of concentration against flow rate, see Notes and Answers. Note that it applies just to the chosen solute output of 750 mosmol/day; moment-to-moment variations in an individual would only follow that curve exactly if the solute output were to stay constant at that value. (Yet Figure 6.3 implies an increase in the urea component with increasing urine flow.)

The maximum concentration of human urine is close to the value calculated for question 6.13.3, i.e. not usually much above 1200 mosmol/l, but sometimes up to 1400 mosmol/l. However, our ability to concentrate urine depends on the solute output. Counterintuitively perhaps, the maximum concentration of urine decreases as the amount of solute excreted increases. Further calculations reveal why this might be.

In the proximal tubules, and again towards the ends of the distal tubules during antidiuresis, the tubular fluid is nearly isosmotic to plasma. Before leaving the kidneys as concentrated urine, fluid must become concentrated from that initial level of 300 mosmol/l by the reabsorption of water in the collecting ducts. The rate of water reabsorption may be estimated from data on the urine such as those used above. For this purpose, the additional solute reabsorption in the collecting ducts can be ignored, since it does not affect the ultimate conclusion.

....

6.13.4 Suppose that the solute output is 600 mosmol/day and that the urine concentration is 1200 mosmol/l. The urine flow rate is thus 0.5 l/day. While still isosmotic with plasma at 300 mosmol/l, the volume of fluid necessary to contain that solute output (still within the renal cortex) must be (600 mosmol/day)/(300 mosmol/l) = 2 l/day. What is the rate of subsequent fluid reabsorption?

6.13.5 On the basis of similar calculations, what would the equivalent rate of reabsorption have to be if the solute output were doubled to 1200 mosmol/day and if the final concentration of the urine were again 1200 mosmol/l?

....

With so much extra water to be reabsorbed into the medullary interstitium it is not surprising that the concentrating ability of the kidneys should be reduced. With the doubled solute output (1200 mosmol/day), the maximum concentration would in fact fall to about 1000 mosmol/l.

6.13.6 What would then be the rate of urine production?

Comparing this rate with the 0.5 l/day of question 6.13.4, we see that, under conditions of maximum antidiuresis, an increase in solute output leads to an increase in the flow of urine. In more extreme cases the phenomenon is called 'osmotic diuresis'. It is strikingly shown when glucose is excreted in diabetes mellitus (Section 6.4). Sucrose and mannitol can be administered as osmotic diuretics; like inulin, these are filtered from the plasma, but not reabsorbed.

Following the same line of argument, it is evident that a person short of water (in the desert, or adrift in a lifeboat) should avoid increasing the excreted solute load. From this point of view, it is better to eat carbohydrate than protein, since the latter yields urea. (Note, however, that some urea is required for the medullary countercurrent mechanism to function.) It is well known that one should not drink sea water in this situation because of its high salt content.

6.13.7 Referring back to questions 6.13.5 and 6.13.6, let us suppose that the extra solute output of 600 mosmol/day in question 6.13.6 was due to the salts of sea water, incautiously drunk by the individual in question. Then the extra solute would correspond to 600 ml of sea water, since the concentration of this is about 1000 mosmol/l. We have already calculated that the rate of urine production would increase from 0.5 l/day to about 1.2 l/day. Would the extra loss of water in urine be more or less than the volume of sea water ingested, and by how much?

The house mouse and numerous desert rodents are much better adapted to conserving water than we are. Thus, the Australian hopping mouse (*Notomys*) can concentrate its urine to 9000 mosmol/l.

6.13.8 Consider again an individual as in question 6.13.4, producing 0.5 l of urine per day at 1200 mosmol/l, with a daily solute output of 600 mosmol. If the urine could be concentrated to 9000 mosmol/l, what would be the daily saving in water?

Would such a saving be worthwhile in ourselves?

We turn now to the extremes of diuresis. What are the highest rates of urine flow and lowest concentrations of urine? It is hard to find precise figures for these. One reason is that the extremes must depend on circumstances, including solute output, but it is also true that most textbooks pay much more attention to antidiuresis. Maximum urine volumes in a day are sometimes to be found, given for example as '15–20 l' or '20–25 l'. Minimum concentrations may be given as '30 mosmol/l or less', 'about 50 mosmol/l' or '50–75 mosmol/l'. What does this inconsistency mean – that the matter has not been adequately explored, or that the minimum concentration depends on circumstances? Short of seeking original data or other accounts, we can hardly resolve that, but we can try putting pairs of these figures together to see whether the implied outputs of solute are likely. For this purpose, let us take the highest of those flow rates (25 l/day) and the low concentrations of 30 and 50 mosmol/l. It seems reasonable that the extremes of flow and dilution should go together, although it should be kept in mind that the figures do come from different sources.

6.13.9 How much solute is excreted per day, in mosmol, when urine flows at 25 l/day and its concentration averages (a) 30 mosmol/l and (b) 50 mosmol/l?

In question 6.12.1, normal values for solute output were estimated as 544–1270 mosmol/day. Both of the values just calculated lie within that range, although the second is rather high to be seen as representative. What we may conclude from these results is that two figures may be coupled in our minds as extremes without obvious inconsistency – a maximum flow rate of about 25 l/day (= 17 ml/min) and a minimum concentration of about 30 mosmol/l. (As a matter of fact, this pairing of rate and concentration has been recorded.) It should not be concluded that this pairing of extremes must apply to all water-loaded individuals under all circumstances, or even that it has to be common. Indeed, the minimum dilution of the urine following considerable water loading is often above 50 mosmol/l. Extreme values are more likely to be sustainable over minutes than over whole days.

6.14 Beer drinker's hyponatraemia

The fact that there is a maximum dilution of urine (Section 6.13) has important implications for those who drink large amounts of fluid, but who have only a small solute load to excrete. Beer has a low salt content and its dedicated drinkers often tend to forgo the normal sources of nutrition and so have little salt and other solute to excrete. This both limits the excretion of water and leads, by a general dilution of body fluids, to hyponatraemia (low plasma sodium). After subsisting for a time on beer, patients are sometimes admitted to hospital suffering from debility and dizziness, and with plasma sodium concentrations even below 110 mmol/l. To explore how this happens, we return to equation 6.9:

$$\text{urine concentration in mosmol/l} = \frac{\text{solute output in mosmol/day}}{\text{urine flow in l/day}}. \quad (6.9)$$

The normal urinary output of solute is about 600–1200 mosmol/day (Section 6.12); let us assume here that in a particular beer drinker it is only 240 mosmol/day. Let us also suppose that this person has 6 l of water to dispose of daily as urine (as compared with a more typical 1–2 l/day). The average urine concentration through the day may now be calculated.

6.14.1 If a beer drinker excretes solute at a rate of 240 mosmol/day with a urine flow of 6 l/day, what is the average concentration of the urine in mosmol/l?

Even though the rate of urine flow is well below the 'normal' maximal rate of 20–25 l/day, the calculated concentration is very low (cf. Section 6.13). To understand how water retention would occur in such an individual, one has only to postulate that urinary dilution is not actually as effective as this over the whole day. If the mean concentration were, say, 60 mosmol/l, the volume of urine produced in a day would be (240 mosmol/day)/(60 mosmol/l) = 4 l, leaving 2 l unexcreted. Actual measurements show that, on admission to hospital, the concentration of the urine can exceed that value of 60 mosmol/l.

An obvious remedy for the water retention is to increase the intake of salt and hence the amount of solute available for excretion. Salted crisps would obviously help, but some drinkers salt their beer.

Compulsive water drinking, as is seen in some psychotic patients, also leads to hyponatraemia. There is often a significant impairment of urinary dilution associated with the hyponatraemia.

6.15 The medullary countercurrent mechanism in antidiuresis – applying the principle of mass balance

Fluids flow into and out of the renal medulla within the vasa recta, loops of Henle and collecting ducts. The total amount of fluid or water entering the medulla must equal the total amount leaving, and, in the long term, this is true of the solutes also. This obvious point exemplifies the principle of mass balance that was applied earlier in relation to renal clearance (equations 1.2 and 6.3), the Fick principle (equation 1.3) and the circulation of blood (Preface and Section 4.6). Here the principle is used to explore the medullary countercurrent mechanism and, in particular, the role of the vasa recta.

Details of this mechanism are to be found in most physiology textbooks and need not be repeated here, especially as many aspects do not now concern us. Very much in outline, urine becomes concentrated in antidiuresis as follows. As tubular fluid flows through the loops of Henle, more solute than water is reabsorbed into the medullary interstitium (the water being reabsorbed in the descending limb, the solute – mainly NaCl – in the ascending limb). This raises the osmotic pressure in the medullary interstitium. Fluid leaving the loops of Henle with a lowered osmolarity flows into the distal convoluted tubules where yet more water and solute are reabsorbed and where the osmolarity returns to that of blood in general (ca. 300 mosmol/l). The tubular fluid passes next to the collecting ducts (with many nephrons converging on each collecting duct). The fluid passing down the medullary collecting ducts becomes concentrated and reduced in volume because, in antidiuresis, their walls are permeable to water and because the osmotic pressure of the surrounding interstitial fluid has been raised by the loops of Henle and by further solute reabsorption in the collecting ducts. The solutes include notably sodium, chloride and urea, but we are not concerned here with distinguishing one solute from another.

Within the medulla, blood flows through the network of vasa recta, first towards the papilla tips and then back to the cortex. A major role of this blood flow in the countercurrent concentrating mechanism is to remove excess solute and water from the medulla, but too brisk a flow would dissipate the osmotic gradient set up by the loops of Henle. Accordingly, much less blood flows to the medulla than to the cortex and most of it travels only into the outer medulla before turning back.

Many textbook descriptions include little on the role of the vasa recta and on flow rates within these blood vessels and the loops of Henle. Here we attempt a quantitative approach, drawing only on the sort of information that is typically available from elementary accounts. It is hardly to be expected that these would include all quantitative details of flow rates and concentrations. Indeed, all of these are variable and, for practical reasons, have not been fully explored in relation to human kidneys. Therefore, in the calculations that follow, it has sometimes been necessary to choose figures somewhat arbitrarily, though within the limits of what is reasonable. As a result, the exact numerical answers are less important than the general conclusions. As a didactic point, we see once again that calculations can be revealing even when they are based on incomplete data.

This section constitutes the longest sustained argument in the book. Partly to avoid complicating it further, a minor simplification is adopted that affects the numerical values rather little and the general conclusions not at all. In the context of osmotic concentrations, the fact is ignored that water makes up only about 80% of the blood. The units used are mosmol/l rather than mosmol/kg water (Section 7.10). All figures refer to a pair of kidneys.

As a preliminary to focusing on the medulla, it may help to apply the principle of mass balance to the simpler case of the whole kidneys, and to seek an answer to the following question: how much would one expect the blood to be diluted as a result of passing through the kidneys in antidiuresis? Suppose that the kidneys are producing urine, in antidiuresis, at a rate of 0.5 ml/min. The renal blood flow is a typical 1200 ml/min, say, and the osmotic concentrations of the afferent blood and the urine are 300.0 and 1200 mosmol/l respectively.

Two points may be made about these figures. First, the precision of '300.0' is necessary here, as the calculations will show, but it must not be confused with accuracy; 300 is just a round number value. Second, both the blood flow and the osmolarity of the urine are numerically '1200'; therefore, to avoid confusion, it is especially sensible to spell out the units throughout the calculations.

Now for the calculations. The amount of solute entering the kidneys in the blood and leaving in urine are respectively [300 mosmol/l × 1200 ml/min ÷ 1000 ml/l] = 360 mosmol/min and [1200 mosmol/l × 0.5 ml/min ÷ 1000 ml/l] = 0.6 mosmol/min. (The flow of slightly hyperosmotic lymph from the kidneys is slow enough not to be significant in this context.) According to these figures, (360 − 0.6) = 359.4 milliosmol leaves the kidneys each minute in (1200 ml − 0.5 ml) = 1199.5 ml of blood.

6.15.1 For comparison with the concentration in the afferent blood, what is the concentration of the blood leaving the kidneys (in mosmol/l to one decimal place)?

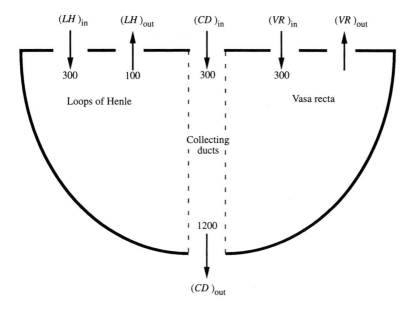

Fig. 6.5. Diagram of the renal medulla showing the inward and outward flows of fluid. The numbers show solute concentrations in mosmol/l and the symbols stand for rates of flow as used in the mass-balance analysis: $(LH)_{in}$ and $(LH)_{out}$, into and out of the medulla via loops of Henle; $(CD)_{in}$ and $(CD)_{out}$, in and out via collecting ducts; $(VR)_{in}$ and $(VR)_{out}$, in and out via vasa recta.

Figure 6.5 shows just the fluids entering and leaving the medulla. Solute concentrations are given for most of these, and symbols representing fluid flow rates (for the pair of kidneys) are also shown. The transition between cortex and medulla is defined here as the limit of the interstitial osmotic gradient. Symbols are defined in the caption, with VR, LH and CD standing for vasa recta, loops of Henle and collecting ducts, respectively. Subscripts indicate whether flow is into, or out of, the medulla.

Applying the principle of mass balance just to fluid flow, we have:

$$(VR)_{in} + (LH)_{in} + (CD)_{in} = (VR)_{out} + (LH)_{out} + (CD)_{out}. \qquad (6.10)$$

As to the numerical values applicable in antidiuresis, the two best known are the rate of flow from the collecting ducts, $(CD)_{out}$, since this is the rate of urine production, and the rate of flow into the loops of Henle from the proximal convoluted tubules. The rate of urine production will be taken again as 0.5 ml/min. The rate of flow into the loops of Henle can be taken as the GFR, say

120 ml/min, less 67–80%, these being textbook figures for the percentage reabsorption of fluid in the proximal convoluted tubules. From within the resulting range of 24–40 ml/min let us choose for $(LH)_{in}$ a round 30 ml/min.

Now for $(CD)_{in}$. If the only process occurring in the collecting ducts were the osmotic extraction of water, the principle would apply that as the volume decreased, the concentration would increase in proportion. Thus, $(CD)_{in}/(CD)_{out}$ would be calculable as the ratio of the concentrations of the fluids leaving and entering the collecting ducts, i.e. (1200 mosmol/l)/(300 mosmol/l) = 4 (cf. question 6.13.4). Since the chosen value of $(CD)_{out}$ is 0.5 ml/min, $(CD)_{in}$ would be $4 \times 0.5 = 2$ ml/min. However, it is an important fact that some urea and salt are reabsorbed from the collecting ducts, so that $(CD)_{in}$ must exceed 2 ml/min. Let us therefore take $(CD)_{in}$ somewhat arbitrarily, and knowing no better, as 3 ml/min. The choice is not very critical.

Similar reasoning may be applied to the loops of Henle on the assumption that their contents are concentrated in the descending limb mainly by osmotic reabsorption of water (which is true in the rabbit, but less so in the rat) and that no water is subsequently gained or lost in the ascending limb (the low permeability of which to water is so important to the countercurrent mechanism). In the longest loops extending all the way to the tip of the papilla, there would the same fourfold increase in concentration as in the collecting ducts. If all loops extended to the tip of the papilla, then $(LH)_{in}/(LH)_{out}$ would equal 4 and, with $(LH)_{in} = 30$ ml/min, $(LH)_{out}$ would be (30 ml/min)/4 = 7.5 ml/min. However, the ratio must be lower than 4 for the greater number of shorter loops. Accordingly, a more realistic average value for $(LH)_{out}$ must be somewhere between 7.5 and 30 ml/min. Let us choose a somewhat arbitrary 10 ml/min.

6.15.2 With $(LH)_{in} = 30$ ml/min and $(LH)_{out} = 10$ ml/min, how much water is reabsorbed in the loops of Henle as a percentage of the chosen GFR of 120 ml/min?

To summarize, we now have the following chosen values for a representative pair of human kidneys in antidiuresis:

$(LH)_{in}$ 30 ml/min	$(LH)_{out}$ 10 ml/min
$(CD)_{in}$ 3 ml/min	$(CD)_{out}$ 0.5 ml/min.

Numerical values for the terms $(LH)_{in}$ and $(CD)_{in}$ are used only in the following question.

6.15.3 On the basis of equation 6.10 and these figures, what is $[(VR)_{out} - (VR)_{in}]$?

The accuracy of the answer is dependent mainly on that of $[(LH)_{in} - (LH)_{out}]$. One may be uneasy about the accuracy of the answer, but its sign must be correct. In other words, more blood must leave the medulla than enters it, having taken up excess water in the vasa recta. This much is evident from equation 6.10, given simply that $(LH)_{in}$ exceeds $(LH)_{out}$ and that $(CD)_{in}$ exceeds $(CD)_{out}$.

There is a problem in comparing $[(VR)_{out} - (VR)_{in}]$ with either of its components, for medullary blood flow is not easily measured. Reflecting the variety of measurements made on experimental animals, textbooks give values for medullary blood flow that range from '0.7%' to '10–15%' of the total renal blood flow, i.e. 8–180 ml/min if the blood flow is taken again as 1200 ml/min. Clearly the calculated value of $[(VR)_{out} - (VR)_{in}]$ is substantial compared with all those values. This means, incidentally, that 'medullary blood flow' depends on whether it is defined as $(VR)_{out}$ or as $(VR)_{in}$.

Referring again to Figure 6.5, the rates of solute movement may now be calculated as products of flow rates and concentrations. All three fluids entering the medulla do so at the same concentration, taken here as 300 mosmol/l. For the osmolarity of the urine in antidiuresis, a figure of 1200 mosmol/l has already been chosen, while another common textbook figure suffices for the dilute fluid leaving the loops of Henle, namely 100 mosmol/l (obviously a round number, and unlikely to be the same for all loops). What we do not know (or so it is assumed for the purpose of the exercise) is the osmolarity of the blood leaving the medulla in the vasa recta. We return to this later, but must first estimate the rate at which solute leaves the medulla in this blood, denoted S. Equating the amounts of solute entering and leaving the medulla, we have:

$$300\{(VR)_{in} + (LH)_{in} + (CD)_{in}\}$$
$$= S + 100(LH)_{out} + 1200(CD)_{out}. \qquad (6.11)$$

The units are nmol/min because rates of solute movement are obtained as products of fluid flow (ml/min) and concentration (mmol/l); the thousand-fold discrepancy between ml and l is taken care of by replacing 'mmol' with 'nmol'. This equation may be combined with equation 6.10 in such a way as to remove three of the flow rates. Thus, if equation 6.10 is multiplied by 300 throughout, its left-hand side becomes the same as that of equation 6.11 and the right-hand side of one then equals the right-hand side of the other. Therefore:

$$S = 300(VR)_{out} + (LH)_{out}\{300 - 100\} - (CD)_{out}\{1200 - 300\}. \quad (6.12)$$

Since $(LH)_{out} = 10$ ml/min, $(CD)_{out} = 0.5$ ml/min and $(VR)_{out} = [(VR)_{in} + 22.5]$ ml/min:

$$S = 300(VR)_{in} + 8300. \quad (6.13)$$

Because $300(VR)_{in}$ is the rate at which solute flows, in the blood, into the vasa recta, S, the rate at which solute leaves, is greater than that by 8300 nosmol/min.

This figure of 8300 is subject to uncertainties in $[(VR)_{out} - (VR)_{in}]$, taken as 22.5 ml/min, and in $(LH)_{out}$, whereas $(CD)_{out}\{1200 - 300\}$ involves only pre-determined quantities.

6.15.4 Which of these terms contributes most to the '8300 nosmol/min': $300[(VR)_{out} - (VR)_{in}]$, $(LH)_{out}\{300 - 100\}$ or $(CD)_{out}\{1200 - 300\}$?

To proceed further, we may now consider the solute and water movements across the walls of an average vas rectum capillary loop, this being defined as the route taken within the medulla by a typical nanolitre of blood. The objective is to estimate $(VR)_{in}$, or rather, since there are many unknown quantities involved, to obtain for it a minimum value. Figure 6.6 is a diagram of our average capillary loop, showing net water movements as solid arrows and net solute movements as broken arrows. (Remember that blood volumes are equated here with volumes of blood water.) The rate of blood flow at the bottom of the loop is reduced by osmotic extraction of water in the descending limb, at rate w, to $[(VR)_{in} - w]$. The corresponding rate of solute flow is $[300(VR)_{in} + s]$. The osmolarity of the blood in that region is increased from its previous value of 300 mosmol/l to a new value denoted C. Therefore, the rate of solute flow at the bottom of the loop is $C[(VR)_{in} - w]$. Equating the two expressions for solute flow, we have:

$$C[(VR)_{in} - w] = 300(VR)_{in} + s. \quad (6.14)$$

A numerical value for C is required now. The maximum osmolarity of the blood, deep in the medulla, is 1200 mosmol/l, matching the urine concentration in this model, but in blood flowing only a short way into the medulla

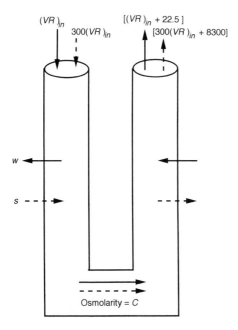

Fig. 6.6. This diagram shows a single capillary loop representing all the networks of vasa recta in the two kidneys. The arrows indicate net movements of water and solutes along the loop and through its walls: solid arrows, movements of water (rates in ml/min); dashed arrows, movements of solute (rates in nosmol/min).

the maximum osmolarity must be closer to 300 mosmol/l. An estimate of the average should take account of the fact that much more blood flows through the outer medulla than through the inner medulla. However, in seeking a minimum estimate of $(VR)_{in}$, we should play safe and choose a value for C that is certainly too high, say $C = 750$ mosmol/l, which is midway between 300 and 1200 mosmol/l. Equation 6.14 accordingly becomes:

$$(VR)_{in} = 1.67w + 0.0022s. \tag{6.15}$$

(The '1.67' is dimensionless; the '0.0022' is in ml/nosmol.) Although we know neither w nor s, we can still seek a minimum estimate of $(VR)_{in}$. The net rate at which solute is added to the blood in the vasa recta was estimated from equation 6.13 as 8300 nosmol/min. Since there is net addition of solute in the descending limb and net reabsorption in the ascending limb, this rate can be taken as the lowest possible value of s, corresponding to zero reabsorption in the ascending limb. As for w, water is certainly drawn out of the descending limb by osmosis, but, without knowing how much, we must, in seeking a minimum estimate of $(VR)_{in}$, take w as zero.

6.15.5 What then is the minimum estimate of $(VR)_{in}$?

Although this is definitely an underestimate, it is higher than the lowest of the values given above that were based on textbook percentages, i.e. 0.7% of 120 ml/min = 8 ml/min. Therefore, for the chosen conditions, that cannot be correct.

6.15.6 What is the corresponding estimate of $(VR)_{out}$?

To estimate the osmolarity of the blood flowing out of the medulla, S is also needed. According to equation 6.13, S is $[300(VR)_{in} + 8300] = [300 \times 18.3 + 8300] = 13,800$ nosmol/min.

6.15.7 What is the osmolarity of the blood flowing out of the medulla?

This is considerably greater than the osmolarity of afferent blood, but is the estimate much inflated by the underestimation of $(VR)_{in}$? To test this, we may choose a higher value for $(VR)_{in}$ and recalculate both $(VR)_{out}$ and S.

6.15.8 What is the osmolarity of the blood flowing out of the medulla if $(VR)_{in}$ is taken as 100 ml/min?

In keeping with the spirit of this book, the countercurrent mechanism has been approached through a sequence of simple calculations, but it is also possible to link the same equations on a computer spreadsheet. Then one may explore, both more easily and systematically, the effects of varying each of the uncertain quantities.

6.16 Renal mitochondria: an exercise involving allometry

The proximal tubules of the kidneys, where most of the fluid and solute reabsorption occurs, contain large numbers of mitochondria. Since these are the site of oxidative metabolism, their abundance should relate to the rate of oxygen consumption. The problem to be addressed here is as follows: since the maximum rate in skeletal muscle is believed to be about 3–6 ml/min per g of mitochondria, can we, for comparison, estimate the equivalent rate for renal mitochondria? Through limitations of data, we explore the matter in relation to the kidneys as a whole, drawing on data for laboratory rats. This provides an opportunity to make practical use of an allometric equation of the sort introduced in Section 1.4 and Chapter 3, that is to say one of the many established equations relating physiological variables to body mass in mammals of widely varying size. References are given in Notes and Answers.

One of the required quantities is the mitochondrial content of the kidneys. I have not seen a figure for human kidneys, but in rats the content has been estimated as 18.2%. Although that is given in terms of volume, it should not be too inaccurate to take the figure as 0.18 g/g (assuming that the densities of mitochondria and of whole kidneys are similar). The percentage was determined in rats of about 0.35 kg. The corresponding volume of a kidney is given as 1.32 ml, so that the mass of a pair would be close to 2.8 g (assuming a density slightly above 1 g/ml).

6.16.1 If 0.18 g of mitochondria is contained in 1 g of kidney, what mass is contained in 2.8 g of kidney?

Only the corresponding rate of oxygen consumption is needed now, but it is not to hand. However, we may estimate it from (1) the GFR and (2) the relationship, already explored in Section 6.8, between GFR and oxygen consumption.

It has been found for a wide range of mammals that GFR is related to body mass in kg, M, through the following approximate allometric equation:

$$\text{GFR (in ml/min)} = 5.36\ M^{0.721}. \tag{6.16}$$

Rat data conform quite well to this.

6.16.2 What is the likely GFR of the above 0.35-kg rats?

In order to relate this answer to oxygen consumption we could work through calculations similar to those in Sections 6.6 and 6.8. However, the sodium concentrations in rat and human plasma are very similar, and an answer can be reached more simply. The human GFR was taken above as 125 ml/min.

For question 6.8.7 the oxygen consumption of the kidneys was taken as 16 ml/min. Thus, the oxygen consumption divided by the GFR is (16 ml O_2/min)/(125 ml fluid/min) = 0.128 ml O_2/ml fluid. If this can be taken as valid for rats, we can now calculate the oxygen consumption of the rat kidney as its GFR (see question 6.16.2) multiplied by that ratio, i.e. (2.5 ml fluid/min) × (0.128 ml O_2/ml fluid) = 0.32 ml O_2/min. (It may not be exactly valid to apply the same ratio to rats as to humans, if only because part of the renal oxygen consumption is not linked to GFR. However, this is a small part, and, like GFR, is likely to be proportionally greater in the smaller mammal.)

6.16.3 If a renal oxygen consumption of 0.32 ml O_2/min is associated in a rat with 0.50 g of mitochondria (as calculated for question 6.16.1), what is the oxygen consumption per g of mitochondria?

The point of these calculations is to compare this answer with the maximum rate in skeletal muscle, i.e. about 3–6 ml O_2/min per g of mitochondria. There is clearly a large discrepancy. In trying to explain it, one should bear in mind that figures have been brought together here from varied sources that may not be strictly compatible. Moreover, some allowance should be made for a 'safety margin' to cope with a higher GFR. If, however, the general conclusion is correct, i.e. that renal mitochondria consume oxygen much more slowly than do mitochondria in fully active muscle, then it is reasonable to ask why the kidneys do not contain fewer, yet more active, mitochondria. In the proximal tubules, most of the mitochondria lie close to the highly invaginated basolateral cell membranes where active sodium transport occurs. Perhaps the abundance of mitochondria there has to do with the need for short diffusion distances between mitochondria and sodium pumps.

According to the above figures, a 0.35-kg rat has 2.8 g of kidneys, i.e. 8 g kidney per kg body mass. A 70-kg person may have kidneys totalling, say, 300 g, so that the equivalent ratio is 4.3 g/kg, i.e. about half the rat value. If the respective GFRs are 2.5 ml/min and 125 ml/min, then the ratios of GFR to kidney mass are 2.5/2.8 = 0.9 ml/g min (rat) and 125/300 = 0.4 ml/g min (human). Thus, compared with us, rats have higher filtration rates for a given mass of kidney, as well as relatively larger kidneys. Recall that small mammals tend to have higher rates of resting metabolism than large mammals and that GFR and basal metabolic rate scale in similar manner with body mass, i.e. with similar exponents (Section 3.10). Although excretion rates for various substances must increase with dietary intake, and hence with metabolic rate, the exact quantitative link between GFR and body mass (equation 6.16) seems yet to be explained.

Another allometric equation relates kidney mass to body mass (M):

$$\text{kidney mass (g)} = 7.3M^{0.85}. \tag{6.17}$$

Let us see how well the human and rat kidney masses used earlier conform.

6.16.4 According to equation 6.17, what are the likely masses of kidney to be found in mammals of (a) 0.35 kg, and (b) 70 kg?

The masses given above are 2.8 g and 300 g respectively. Given that such allometric equations are never expected to be exactly predictive, the agreement is reasonable.

7 Body fluids

Renal function and acid–base balance are treated in separate chapters. Here, the first questions concern effects of ingesting water and potassium, and the movements of ions between extracellular fluid and both cells (Section 7.3) and bone mineral (Section 7.4). The amount of bone mineral in the body is discussed in Section 7.5.

Section 7.6, on the important principle of electroneutrality then leads us more in the direction of physical chemistry – to Donnan equilibria and colloid osmotic pressure (Sections 7.7 and 7.8). Proteins may influence ionic concentrations not only through the Donnan effect, but, when the concentrations are in molar rather than molal terms, by simple dilution; the distinctions between molarity and molality and between osmolarity and osmolality are explored in Sections 7.9 and 7.10, with a view to making the distinctions less tiresome for those who need to acknowledge them.

Sections 7.11 and 7.12 are about things more commonly found under such headings as 'excitable tissues' or 'nerve and muscle' (Chapter 9), these being the relationships amongst membrane potentials and transmembrane gradients of sodium and potassium.

Some aspects of water balance are discussed elsewhere – metabolic water (Section 3.14), water loss in expired air (Section 5.6) and urine (Chapter 6). Osmoles and osmotic pressures are treated briefly in Notes and Answers.

7.1 The sensitivity of hypothalamic osmoreceptors

As noted earlier, the brochures for at least two kinds of commercial osmometer give their accuracy as 'within 2 mosmol/kg water'. How do our hypothalamic osmoreceptors compare?

7.1.1 The body of a man contains 49.5 l of water. He drinks 0.5 l of water. Suppose that all this water is absorbed, at first with negligible excretion.

What are (a) the percentage decrease in osmotic concentration of the body fluids and (b) (given an initial concentration of 300 mosmol/kg water) the absolute decrease?

Half a litre of water is usually enough to produce a small diuresis in someone who is in a normal (neutral) state of water balance, and the typical diuretic response suggests that our hypothalamic osmoreceptors perform at least as well as the above osmometers. Indeed, a mere 1% fall in osmotic concentration is generally enough to effect a significant decrease in the secretion of antidiuretic hormone. Note, however, that the argument is not watertight as it stands, for the arterial blood supplying the osmoreceptors, and even more so the blood leaving the gastrointestinal tract, would temporarily be more diluted than the other body fluids. For a full analysis of the situation, the time courses of gastrointestinal absorption, mixing and diuresis should also be taken into account. The osmoreceptors vary in their sensitivity to changes in osmotic pressure, some responding to a change as small as 0.2% and others only to about 1.5%.

7.2 Cells as 'buffers' of extracellular potassium

Food is mostly derived from plant and animal cells, and cells contain much potassium. Let us consider the effects of ingesting a specific amount of potassium, namely 30 mmol. This could be contained in, say, 300 g of skeletal muscle, hot-pot or banana, or 200 g of raw potatoes.

7.2.1 Suppose that this 30 mmol of potassium were to be absorbed into 15 l of extracellular fluid without any of it being excreted or entering other body compartments. By how much would the extracellular concentration rise? What would the concentration then be if it were previously 4.5 mmol/l?

Ventricular fibrillation may occur when plasma potassium exceeds about 6 mmol/l – although clearly not when it happens, as it often does, as the result of exercise! Considering that a person might not only absorb the above amount of ingested potassium, but also take exercise, the initial assumption that all the ingested potassium stays in the extracellular fluid will not do. Indeed, the concentration in plasma is known not to rise as much as the calculation indicates. An obvious thought is that the excess potassium is as rapidly excreted by the kidneys as it is absorbed from the gut, but it is also a possibility that some of the potassium temporarily enters the cells. In fact the latter is an important mechanism of potassium homeostasis and although our calculation does not suffice to prove it, there is plenty of experimental evidence that the cells in general do act to 'buffer' extracellular potassium. Hormones such as insulin and catecholamines are involved in this mechanism.

Experimentally, subjects have taken seven or more times the above 30 mmol of potassium during the course of a day, with plasma potassium increasing less than 0.3 mmol/l.

At this point it would be satisfying to calculate the rate at which the kidneys might excrete the ingested potassium and relate that to the rate of absorption from the gastrointestinal tract. Unfortunately the latter rate has not been well studied. It does seem, however, that absorption from the gut would take long enough (a few hours) for renal excretion to be proceeding significantly.

In considering the role of the cells in buffering extracellular potassium, it helps to be aware of the distribution of potassium within the body. As round numbers, the concentration of potassium in the cells averages 150 mmol/l water and the concentration in the extracellular fluid is 5 mmol/l water. The volumes of intracellular and extracellular water in a particular individual might be about 30 l and 15 l respectively.

....

7.2.2 On the basis of those figures, how much potassium is there in (a) all the cells of the body and (b) the extracellular fluid?

7.2.3 Assuming initial conditions as just specified, if all the above 30 mmol of ingested potassium were to enter the cells, what would be the percentage increase in average intracellular concentration?

....

Would such a change be experimentally detectable? Is it likely that the change would influence cell function significantly?

7.3 Assessing movements of sodium between body compartments – a practical difficulty

The problem discussed here is mainly relevant to experimental studies, but is more generally worth pursuing as an exercise in the application of both elementary ideas and usefully remembered quantities.

Suppose that there is net movement of a substance between tissues and extracellular fluid within the whole body and that this movement needs to be quantified experimentally. Often it suffices to note how the concentration changes in the blood plasma. This is the case with, say, potassium or glucose, but not, as will be seen, with sodium.

To consider a particular example, suppose that sodium and potassium move with bicarbonate from cells to extracellular fluid and that they do so in a ratio of 3 sodium ions to 1 potassium ion and 4 bicarbonate ions. This situation and these ratios have been chosen because similar ionic shifts are thought to occur in response to respiratory acidosis (Section 8.8), but the concern here is not particularly with acid–base balance as such. The point is that the movement of the sodium, potassium and bicarbonate may be expected to lead to osmotic movement of water in the same direction. Whether the concentration of sodium in the extracellular fluid rises or falls depends on the relative amounts of sodium and water that enter it.

Assuming that osmotic equilibrium is restored after these movements of ions and water, we may think of the whole process as the movement of an isosmotic solution out of the cells. Let us take the osmolality of this solution, of the cells, and of the extracellular fluid as 300 mosmol/kg water.

Suppose now that 3 mmol of sodium moves from the cells into each kilogram of extracellular water. Given that these sodium ions are accompanied by 5 mmol of other ions (1 mmol of potassium plus 4 mmol of bicarbonate), a total of 8 mosmol of assorted ions enters each kilogram of extracellular water. (Here, with trivial consequence, we equate moles and osmoles.) We are assuming that these solutes, accompanied by water, move effectively as a solution containing 300 mosmol/kg water and this means that the 8 mosmol takes with it 27 g of water (since 8 mosmol/27 g = 300 mosmol/1000 g). This, then, is the amount of water accompanying the 3 mmol of sodium into each kilogram of extracellular water.

7.3.1 If 3 mmol of sodium moves from the cells into each kilogram of extracellular water accompanied by 27 g of water, and if the initial extracellular sodium concentration is 150 mmol/kg water, what is the final concentration of sodium?

It is evident that shifts of sodium between extracellular and intracellular fluid cannot be assessed simply on the basis of changing extracellular concentrations – even as regards direction. It is also necessary to know what is happening to the extracellular fluid volume.

The total sodium content of the extracellular fluid is calculable as the product of its volume (or the mass of its water) and the sodium concentration. If both of these can be measured exactly, then the amount of sodium gained or lost in a given situation can be calculated as a difference between two products. The extracellular fluid volume may be estimated as the volume of distribution of a substance, such as inulin, that spreads evenly throughout the volume and that does not enter the cells (see Notes to Chapter 2).

Difficulties remain. The 'inulin space' is subject to error and the method requires corrections for the loss of inulin in the urine. The sodium concentration, best expressed in molal rather than molar terms (Section 7.9), is not always easy to measure to the accuracy called for in question 7.3.1. Moreover, the plasma and interstitial fluids should differ slightly in molal sodium concentration because of the Donnan effect (Section 7.7).

There is another general point to be made, this time concerning homeostasis itself, rather than its study. Whenever sodium moves between cells and extracellular fluid with an equivalent amount of univalent anion (not necessarily accompanied by potassium), then water moves too and the extracellular concentration of sodium hardly alters. The concentration in the extracellular fluid cannot be significantly adjusted just by shifts of sodium plus anion, but must be regulated by other means. It is a different matter if sodium crosses the cell membranes in what is effectively a 1:1 exchange for other solutes, for then there is no net water movement.

7.4 The role of bone mineral in the regulation of extracellular calcium and phosphate

The continual precipitation and dissolution of bone mineral is an important factor in the homeostasis of extracellular calcium and inorganic phosphate. It is modulated by such hormones as parathyroid hormone and calcitonin. Bone mineral seems to consist mainly of four forms of calcium phosphate. Here are their formulae:

$$Ca_9(PO_4)_{4.5}(CO_3)_{1.5}(OH)_{1.5}$$
$$Ca_9Mg(HPO_4)(PO_4)_6$$
$$Ca_{8.5}Na_{1.5}(PO_4)_{4.5}(CO_3)_{2.5}$$
$$Ca_8(HPO_4)_2(PO_4)_{4.5}H_2O.$$

The first three have very low solubilities and are far from being in equilibrium with extracellular fluid. The fourth is octocalcium phosphate. This one is much the most soluble and is believed to be in equilibrium with bone extracellular fluid and close to equilibrium with the rest of the extracellular fluid. The relationship between the solubility product for octocalcium phosphate and the concentrations of the relevant ions is one that can obviously be approached quantitatively, though not very simply, but that is not the intention here. Suffice to say that the solubility product for octocalcium phosphate has evidently been one of the determinants of extracellular fluid composition during the evolution of mammals.

We now consider a different feature of this composition, namely the similarity between the concentrations of inorganic phosphate ($HPO_4 + H_2PO_4$) and of free calcium ions. Both are close to 1.3 mmol/l, so that the ratio of calcium to phosphorus (Ca/P) is close to 1. This fact is not predictable from the solubility relationship, for either concentration could be much higher, provided that the other were correspondingly lower. That ratios very different from 1 are possible in extracellular fluids is illustrated by gastropods and other invertebrate animals. In many of these the calcium concentration exceeds 10 mmol/l, while the concentration of inorganic phosphate is only 0.1–0.3 mmol/l. Invertebrates do not have bone mineral, but may contain other forms of precipitated calcium phosphate.

7.4.1 In the aforementioned four bone salts, what is the range of values of the mole ratio Ca/P?

The point to be made next depends on the similarity of these ratios to the Ca/P ratio in mammalian extracellular fluid. They are not identical, but they are clearly much more similar to each other than they are to the ratios of 30–100 in the aforementioned invertebrates. The following three calculations are about the effects on extracellular fluid of the dissolution and precipitation of bone mineral. Assume for these calculations that there are no exchanges with other body compartments and no change in the extent of binding of calcium to proteins.

7.4.2 Suppose that calcium and phosphate dissolve into mammalian extracellular fluid from bone mineral in a ratio of, say, 1.5 Ca to 1 P – to the extent that the concentration of inorganic phosphate rises from 1.0 mmol/l

to 1.3 mmol/l (30%). What is the percentage rise in free calcium, if the initial value is 1.3 mmol/l?

The changes in the concentrations of calcium and phosphate are proportionately similar. To appreciate the point of this, consider now a hypothetical bony animal where that is not true. Its extracellular concentrations are much as in some invertebrates: free calcium, 10 mmol/l; inorganic phosphate, 0.1 mmol/l. Assume as before that calcium and phosphate dissolve from the bone mineral, or precipitate onto it, in the ratio of 1.5 Ca to 1 P, and that there are no exchanges with other compartments of the body.

7.4.3 If the concentration of inorganic phosphate is again raised by 30%, by what percentage is the concentration of calcium raised this time?

7.4.4 Considering the same hypothetical animal, and the same initial conditions, suppose, very improbably, that all the extracellular phosphate were to precipitate. What would be the final concentration of calcium in the extracellular fluid?

From a comparison of the last three answers, it is evident that bone mineral can 'buffer' both ions effectively in the mammal, but not in the hypothetical animal. In the latter, the mineral could only contribute significantly to the homeostasis of calcium if huge variations in dissolved phosphate were tolerated – or else prevented by additional mechanisms such as renal regulation. The similarity of calcium and phosphate concentrations in mammalian extracellular fluid can thus be seen as advantageous. Whether the advantage helped to mould the evolution of human body fluids is an open question, however. It must also be said that the similarity of calcium and phosphate concentrations also gave rise to a potential problem in homeostasis. Since calcium and phosphate necessarily move together between extracellular fluid and bone mineral, the other mechanisms that regulate them must be integrated accordingly. Note, for example, that when parathyroid hormone acts to raise extracellular calcium, it both mobilizes phosphate from bone mineral and increases the rate at which phosphate is excreted by the kidneys.

7.5 The amounts of calcium and bone in the body

How much of the typical human body consists of bone or skeleton? The typical textbook does not say, nor is it always clear whether published figures refer to living bone or to air-dry museum skeletons, nor whether marrow and other components are included (an uncertainty that often applies to other mammals). What are easier to find in textbooks are indications of the total calcium content of the human body and, since almost all is in bones and teeth, such figures can be used for estimations of bone content. The proportion of calcium in the body varies, not only with skeletal development and mineralization, but also with the body's fat content. However, a 70-kg body can be said to contain about 1 kg of calcium as a round number, or 1.4% (see Notes and Answers). This can be taken as the starting point for calculations based on the calcium content of bone itself. (Since only an approximate, or representative, answer is being sought here for the amount of bone in the body, the teeth may be disregarded.)

Bone, freed of marrow and other connective tissue, contains not only the mineral component, consisting mainly of the calcium phosphates of Section 7.4, but also the organic matrix, cells and extracellular fluid. As its exact composition varies a little from site to site, so too does its calcium content. The latter is usually given in terms of dry, fat-free bone, a typical value being about 26%.

7.5.1 Accordingly, what mass of dry, fat-free bone contains that 1 kg of calcium?

7.5.2 If that amount of bone is present in a 70-kg body, what percentage of the latter consists of dry, fat-free bone?

The calcium content of dry, fat-free bone (i.e. ca. 26%) may be compared with the calcium content of its mineral component. Section 7.4 gives the formulae of four forms of calcium phosphate. The percentage of calcium in each may be calculated (using the atomic masses of all the component atoms), the four values being, in the order in which the substances are presented, 40%, 34%, 36% and 33%. The most abundant of the four forms of calcium phosphate is the first, containing 40% calcium by mass.

7.5.3 If bone mineral consisted entirely of that form, and the body contained 1.4% bone calcium as before, what percentage of the body would consist of bone mineral?

7.6 The principle of electroneutrality

According to this principle, in any solution the total charge on all the anions present is virtually equal to the total charge on all the cations. In other words, the concentration of anions in terms of equivalents is virtually equal to the concentration of cations in the same units. (For each ion in a solution, the number of equivalents present is equal to the number of moles multiplied by the respective valency.) This is a key idea in electrolyte and acid–base physiology, but it is given curiously little emphasis in many elementary accounts and seems sometimes to be forgotten even by research workers. Examples of its application are given below, but first that word 'virtually' must be considered.

The existence of a membrane potential implies that there is a slight imbalance of anions and cations across the cell membrane. When, as is typical, the interior of a cell is negative with respect to the exterior, then there is a slight excess of anions over cations inside. Just how much of a discrepancy there is depends on the magnitude of the membrane potential (V_m, in volts) and membrane capacity (C, in farad/cm²). The charge on each square centimetre of membrane, in coulombs, is the product of these two (1 farad × 1 volt = 1 coulomb). Our concern is more with amounts of ion than with coulombs, but the two are related through the Faraday constant, F. This is 96,490 coulombs/equiv. In terms of mol/cm² of membrane, the discrepancy is given by CV_m/zF, where 'z' is the valency of the ions.

For a variety of cell types, the membrane capacity is close to 10^{-6} farad/cm² (more usually quoted as 1 microfarad/cm²). As a representative (resting) membrane potential, we may take -0.07 volt (-70 mV). For the present purpose the ions may be considered as all univalent (so that $z = 1$ and 1 mol = 1 equiv).

....

7.6.1 For a cell with membrane capacity, C, of 10^{-6} farad/cm² and a (negative) membrane potential of 0.07 volt (70 mV), what is the discrepancy between the concentrations of anions and cations (in mol/cm²), calculated as CV_m/zF?

....

The answer needs now to be expressed in terms of mmol/l, but it then depends on the size and shape of the cell. Consider a length of cylindrical nerve fibre of diameter D and length L, both expressed in cm. The surface area and volume are respectively πDL cm² and $\pi D^2 L/4$ cm³, so that the ratio of surface area to volume is $4/(D\,cm)$. Given the answer to question 7.6.1, the discrepancy in mol/cm³ is 7×10^{-13} mol/cm² $\times 4/(D$ cm$) = 28/D \times 10^{-13}$ mol/cm³. Since 1 mol/cm³ $= (1000$ mmol$)/(0.001$ l$) = 10^6$ mmol/l and 1 cm $= 10^4$ μm, we arrive at the following equation:

$$\text{discrepancy in mmol/l} = 0.028/(\text{diameter in } \mu m). \tag{7.1}$$

7.6.2 What is this discrepancy between anions and cations in (a) a nerve fibre of diameter 7 μm, and (b) a nerve fibre of diameter 1 μm?

Compare these minute quantities with the typical intracellular potassium concentration of about 150 mmol/kg water. Would it be possible to demonstrate the discrepancy by chemical analysis? There are many different kinds of ion present and it is virtually impossible to draw up a detailed and accurate balance sheet for all the anions and cations in a particular cell or population of cells. Among the difficulties involved are those of identifying all the organic substances present and, when analysing a piece of tissue, of making exact allowance for the presence of extracellular fluid.

The ionic movements associated with action potentials are likewise tiny. Indeed their minuteness is important in allowing rapid changes of V_m, and then the re-establishment of ionic gradients with the minimum expenditure of energy. The discrepancy between the total concentrations of anions and cations has not been demonstrated directly. However, the amounts of sodium and potassium entering and leaving nerve fibres during action potentials have been measured; they are similar to the amounts calculated in the manner illustrated above.

Let us now apply the principle of electroneutrality to blood plasma. Suppose that a sample of normal human plasma is analysed and found to contain ions at the following concentrations (mmol/l):

Na   144;   K   4;   free Ca   1;   free Mg   0.5;
Cl   102; HCO_3 28;   lactate   1.

The concentration of protein (which has a net negative charge) is estimated as 18 mequiv/l (Since the number of equivalents of each ion is equal to the number of moles multiplied by the respective valency, 18 mequiv of protein carries the same charge as 18 mmol of univalent anion.)

7.6.3 In mequiv/l, what is the difference between the total measured concentration of cations in this plasma sample and the total estimated concentration of anions?

That the difference is not zero may be partly due to analytical error, but a small discrepancy is to be expected anyway, since some of the minor ionic constituents, such as sulphate and inorganic phosphate, have not been measured.

Suppose now that another sample of blood plasma is analysed in another laboratory. This time it is from a Transylvanian blue-eyed bloater, a fish that has never previously been studied. The analysis, less complete, is as follows (with concentrations in mmol/l):

$$Na \quad 138; \qquad K \qquad 3; \qquad total\ Ca \quad 1.5; \qquad total\ Mg \quad 3;$$
$$Cl \quad 70; \qquad HCO_3 \quad 5.$$

7.6.4 Again in mequiv/l, what is the discrepancy between measured cations and anions? What could it mean?

The example is imaginary, but it is not unlike many analyses that have been published without relevant comment. The discrepancy is large, but its meaning is not clear. Does the plasma contain a higher concentration of (negatively charged) protein or other anion than is usual in human plasma? Are the analyses faulty? What is clear is that there is something to be investigated further. It often pays to check the balance of anions and cations.

Human erythrocytes (in arterial blood) contain sodium, potassium, chloride and bicarbonate at approximately the following concentrations (mmol/kg water):

$$Na\ 18; \quad K\ 135; \quad Cl\ 78; \quad HCO_3\ 16.$$

Free calcium is negligible in the context of charge balance. The concentration of free magnesium is about 0.5 mmol/kg water.

7.6.5 What is the discrepancy between measured anions and cations, in mequiv/kg water? How might it be interpreted?

7.7 Donnan equilibrium

The glomerular filtrate has been described as resembling plasma that is nearly free of its proteins (except that any substances bound to the proteins would be retained with them in the plasma during filtration). More detailed consideration reveals that there must be other minor differences in ionic composition between plasma and filtrate, these being due to the net negative charge on the plasma proteins. This brings us to the topic of the Donnan (or Gibbs–Donnan) equilibrium. This is also relevant to the filtration of fluid through the walls of capillaries generally, and to the experimental dialysis of protein solutions. Although the subject seems sometimes to be over-emphasized in elementary teaching, it can be worth looking at for the use made of the Nernst equation and the principle of electroneutrality. It also has an important bearing on colloid osmotic pressures (Section 7.8).

Consider any two solutions that are separated by a membrane that is permeable to all ions present except proteins. It could, for example, be a cellulose dialysis membrane or, less ideally, capillary endothelium. The permeant ions are allowed to reach diffusion equilibrium. (As is explained below, there is likely to be a difference in osmotic pressure between the two solutions, this being the colloid osmotic pressure. To prevent water movement, there would therefore need to be a small opposing difference in hydrostatic pressure.)

A simple rule relates the various concentrations of diffusible, unbound ions:

$$\frac{[Na]_1}{[Na]_2} = \frac{[Cl]_2}{[Cl]_1} = \frac{\sqrt{[Ca]_1}}{\sqrt{[Ca]_2}} = \frac{\sqrt{[SO_4]_2}}{\sqrt{[SO_4]_2}} \ldots \text{ and so on.} \tag{7.2}$$

The subscripts 1 and 2 refer to the two solutions. Square roots are required for all divalent ions.

If this set of relationships is unfamiliar, note that it may be derived by equating the equilibrium potentials for all ionic species present, except protein, these potentials being given in each case by the Nernst equation. Thus, for sodium and chloride at 37 °C,

$$E_{Na} = 61.5 \log\frac{[Na]_1}{[Na]_2} = E_{Cl} = 61.5 \log\frac{[Cl]_2}{[Cl]_1} \tag{7.3}$$

Now divide through by 61.5 and take antilogarithms and you have the first part of equation 7.2.

For simplicity, let us take solution 1 as containing just NaCl, at concentration C mmol/kg water. Solution 2 contains NaCl too, but it also contains protein at concentration E mequiv/kg water. In accordance with the principle of electroneutrality:

$$[Na]_1 = [Cl]_1 = C \tag{7.4}$$

and

$$[Na]_2 = [Cl]_2 + E. \tag{7.5}$$

From equations 7.2, 7.4 and 7.5:

$$([Cl]_2 + E)[Cl]_2 = C^2. \tag{7.6}$$

Thus (an equation to apply here, but not remember):

$$2[Cl]_2 = (E^2 + 4C^2)^{1/2} - E. \tag{7.7}$$

Consider now a pair of solutions such that C is 150 mmol/kg water and E is 18 mequiv/kg water, values chosen to be like those in human plasma. $[Cl]_2$, the concentration of chloride in the solution with the protein, works out as $\{[18^2 + (4 \times 150^2)]^{1/2} - 18\}/2 = 141.3$ mmol/kg water.

....

7.7.1 According to equation 7.5, what is $[Na]_2$?

7.7.2 As a check, is $[Na]_1/[Na]_2$ equal to $[Cl]_2/[Cl]_1$, in accordance with equation 7.2?

....

The general magnitude of this 'Donnan ratio' for plasma may be worth remembering – close to 1, yet readily distinguishable from it by chemical analysis. Both $([Na]_2 - [Na]_1)$ and $([Cl]_1 - [Cl]_2)$ work out nearly equal to $E/2$ and this is a convenient conclusion too.

In practice, it is not usually fruitful to apply these equations to plasma with the expectation of a precise answer. E is rarely known exactly and even equation 7.2 tends not to fit analytical data, partly because of the specific binding of ions to proteins. (What is more, the difference between binding and Donnan effects can be difficult to separate, even conceptually.)

One other aspect of the Donnan equilibrium is, however, worth quantifying before we go on to look at the relevance of the equilibrium to colloid osmotic pressure. This is the electrical potential between the two solutions. Since both sodium and chloride ions are at equilibrium, the electrical potential is equal to both E_{Na} and E_{Cl} in equation 7.3, this involving the Nernst equation formulated in terms of mV at 37 °C. $[Na]_1/[Na]_2$ and $[Cl]_2/[Cl]_1$ have already been calculated (question 7.7.2).

7.7.3 Solution 2, which contains the (negatively charged) protein, is slightly negative with respect to solution 1. What is the electrical potential difference? (log 0.942 = −0.026)

7.8 Colloid osmotic pressure

The colloid osmotic pressure, or oncotic pressure, of a solution is that part of the total osmotic pressure that is due to colloids. In natural body fluids the colloids are the proteins. The colloid osmotic pressure of human plasma varies between about 20 and 35 mmHg.

It is important to realize how very different this value is from the total osmotic pressure, but because the two are usually expressed in different units the point is often unappreciated. The total osmotic concentrations of the body fluids are not usually given in pressure units, but as osmolalities (see Notes and Answers). As a round number, the osmolality of most human body fluids (cells and extracellular fluid) is 300 mosmol/kg water. As discussed in Notes and Answers, 1 mosmol/kg water exerts an osmotic pressure of 19.3 mmHg at body temperature.

7.8.1 Given that relationship, and assuming an osmolality of 300 mosmol/kg water, what is the osmotic pressure of human body fluids in mmHg?

Compare this with the colloid osmotic pressure of human plasma (20–35 mmHg).

It might be supposed that colloid osmotic pressures would be related to protein concentrations in accordance with the relationship just given for other solutes. In general, colloid osmotic pressure does increase with protein concentration and, for a given concentration in g/l or g/kg water, proteins of low molecular weight (e.g. albumins) do exert higher osmotic pressures than do those of high molecular weight (e.g. globulins). In short, it is roughly true to say that the colloid osmotic pressure increases with the

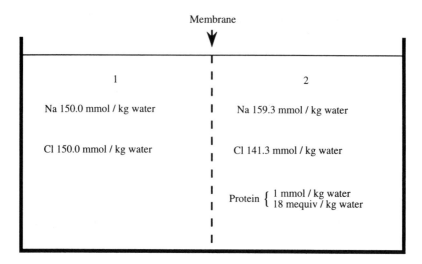

Fig. 7.1. Solutions 1 and 2 are separated by a semipermeable membrane and are in Donnan equilibrium. Sodium and chloride are present at the concentrations shown. Proteins are present in solution 2 only and give rise to a colloid osmotic pressure that tends to draw water from left to right.

concentration of protein in terms of mmol/kg water (doing so more and more steeply as the concentration increases). This may be as much most physiologists need to know on the subject, but it is an easy matter to probe a little further making use of data obtained already in Section 7.7.

Figure 7.1 summarizes the data associated with question 7.7.2. Solutions 1 and 2 are separated by a membrane that is permeable to everything except protein. The solutions are in Donnan equilibrium. Solution 1 contains just NaCl and solution 2 contains protein at a concentration like that in blood plasma.

7.8.2 In which solution is the sum of concentrations ([Na] + [Cl]) greater, the one with the protein (2) or the one without it (1)? By how much is the sum greater?

It is evident that the protein in solution 2 raises the osmotic pressure above that in solution 1 both directly and indirectly. The 'colloid osmotic pressure' of solution 2 is not only due to the direct effect of the proteins themselves, but also to the slight excess of ions that they attract.

To quantify these effects it suffices here to regard 1 mosmol of protein, sodium or chloride as the same as 1 mmol, as in an 'ideal' solution. Accordingly, 1 mmol/kg water of any of these exerts an osmotic pressure of 19.3 mmHg at body temperature (see above).

In the calculation of the sodium and chloride concentrations in solution 2 the concentration of protein had to be expressed in terms of charge, i.e. as 18 mequiv/kg water. It must now be specified in terms of mmol and a reasonable figure to take is 1 mmol/kg water.

7.8.3 In mmol/kg water, how much more solute (i.e protein + sodium + chloride) is there in solution 2 as compared with solution 1?

This difference in total concentration is responsible for the colloid osmotic pressure of solution 2.

7.8.4 What is the answer to question 7.8.3 expressed in mmHg, assuming that 1 mmol/kg water exerts an osmotic pressure of 19.3 mmHg at body temperature?

The calculated value is fairly typical of human plasma, and indeed the colloid osmotic pressure of plasma is due both to the direct effect of the protein and to the asymmetrical distribution of inorganic ions, whether that be across endothelium or across the artificial membrane of a colloid osmometer. Unfortunately real protein solutions do not behave quantitatively in quite the simple way that has so far been implied and for this reason their colloid osmotic pressures are better measured than calculated.

The protein concentration was taken above as 1 mmol/kg water, but more usual units for protein are g/l or g/kg water. For any given protein, the concentration in g/kg water is calculable as the concentration in mmol/kg water multiplied by one-thousandth of the molecular weight.

7.8.5 A solution contains protein at 1 mmol/kg water as above. What would the concentration be in g/kg water if the protein were albumin of molecular weight 68,000?

The protein content of plasma is typically 65–80 g/kg water, with about 60% being albumin. About 35% is globulin. The globulin has a much higher average molecular weight and so contributes proportionately less to colloid osmotic pressure.

7.9 Molar and molal concentrations

Concentrations of solutes in body fluids may be expressed in terms of millimolarity (mmol/l of solution, mM) or millimolality (mmol/kg of water). In extracellular fluid the numerical difference is small enough that for many of the calculations in this book it is unimportant. Moreover, in an elementary course of physiology the difference may well be regarded as a distracting complication that is best ignored. Nevertheless, the matter is important to the interpretation of actual clinical measurements on plasma, and with cytoplasm the difference between the two measures is large.

Let us start by defining a conversion factor, w, such that

$$w = \text{millimolar concentration/millimolal concentration}$$
$$= \frac{\text{mmol/l solution}}{\text{mmol/kg water}}$$
$$= \text{kg water/l solution, or g water/ml solution.} \tag{7.8}$$

Consider now a salt solution that contains proteins at concentration c g/l. Let w have a value of w_0 when c is zero. In the case of human plasma without its proteins w_0 is about 0.99 kg water/l solution. Let the volume occupied by a gram of protein be V ml (V being the 'partial specific volume' of the protein). For most proteins, V is 0.70–0.75 ml/g; for plasma proteins it is about 0.75 ml/g. The various quantities are related thus:

$$w = w_0(1 - Vc/1000), \tag{7.9}$$

or, for blood plasma,

$$w = 0.99(1 - 0.75c/1000). \tag{7.10}$$

The equation is not one to memorize, but the calculating physiologist may find it useful to have it recorded here.

....

7.9.1 What is w for plasma with a (typical) protein concentration of 70 g/l?

7.9.2 If plasma with that protein content contains sodium at 141 mmol/l, what is the concentration of sodium in mmol/kg water?

....

Since 141 mmol/l is within the normal range for sodium of about 137–145 mmol/l, 150 mmol/kg water can be taken as a convenient round-number value for plasma in general.

So far, one component of plasma has been ignored that can be very relevant, namely the lipids. Normal concentrations in adults are about 5–9 g/l. They make no difference to the millimolal concentrations of other substances, but lower their millimolar concentrations by adding to the total volume. In cases of hyperlipaemia, there may therefore be an apparent hyponatraemia when the sodium concentration is recorded in mmol/l – even though the concentration in mmol/kg water is normal. It is the concentration in mmol/kg water that matters so far as cell function is concerned.

In practice, one might wish to effect the conversion from millimolar to millimolal on the basis of two measurements that were not used above, namely the density of the solution and its water content (g water/g solution) as determined by weighing and drying. Then:

$$w = \text{density (g/ml)} \times \text{water content (g/g).} \tag{7.11}$$

In cytoplasm, the difference between millimolar and millimolal concentrations is much greater than in plasma, because of the large amount of substance present that is not water. Those who analyse tissues generally do so on a basis of mass rather than of volume and end up neither with concentrations in mmol/l of cytoplasm, nor with the information necessary for the calculation of these. Some authors of textbooks, on the other hand, do write, loosely, of 'mM' (which is strictly an abbreviation for mmol/l). Since millimolar concentrations are rarely meaningful in this context, we may be content with just a rough calculation to illustrate the difference between units.

For this purpose, a representative value of c is needed – but this time including all the organic components of the cytoplasm and not just the protein. For erythrocytes, a figure of 340 g/l may be familiar as the 'mean corpuscular haemoglobin concentration' (MCHC); c in erythrocytes is only slightly higher – about 360 g/l. This is also about right for some other cells.

....

7.9.3 Treating cytoplasm as if it were like very concentrated blood plasma, and applying equation 7.10, what is the conversion factor, w, if c is 360 g/l?

7.9.4 If a cell contains potassium at a concentration of 150 mmol/kg water, what is the concentration in mmol/l, assuming the value of w just calculated?

7.10 Osmolarity and osmolality

Whereas the mole and millimole are usual units for individual solutes, the osmole and milliosmole, defined in Notes and Answers, are generally only used in the context of total solute concentration, and especially where that relates to osmotic pressure. The difference between osmolarity and osmolality is the same as the difference between molarity and molality; osmolarity is the concentration in osmol/l of fluid and osmolality is the concentration in osmol/kg water. As with molarity and molality, the distinction, simple though it is, can be more troublesome than useful in an elementary course in physiology.

Whilst osmolality is directly measurable in terms of freezing point or vapour pressure, osmolarity cannot be directly measured and is not generally a useful concept. If people tend to speak more of osmolarities, perhaps it is because the word is more euphonious.

One context in which it does arguably make sense to think in terms of osmolarities rather than osmolalities is that of comparing the osmotic concentration of a solution with the millimolar concentrations of the various contributory solutes. One might, for example, add up all the solutes given for the plasma sample of question 7.6.3 to see if they are roughly compatible with normal values of osmolality/osmolarity. Here again are those concentrations (in mmol/l):

Na 144;   K   4;   free Ca 1;   free Mg 0.5;
Cl 102; HCO_3 28;   lactate 1.

The protein concentration can be taken as about 1 mmol/l.

7.10.1 What is the sum of all these concentrations?

A round-number osmolality of 300 mosmol/kg water was used for plasma earlier. Actual values are usually in the range of 290–300 mosmol/kg water. Considering that the units are different, the match could be regarded as reasonable, especially since other solutes, such as glucose and urea, are not included in the total for question 7.10.1. For a more accurate comparison, corrections are needed for (a) the difference between osmolarity and osmolality, and (b) the non-ideal behaviour of solutions, i.e. the fact that 1 mmol of solute is generally less than 1 mosmol.

At this point one may be tempted to skip the tedium of making such corrections for what is only a hypothetical example – and to skip therefore the next calculation. The reward for reading on is the demonstration that calculation happens often to be unnecessary.

To convert the above answer to mmol/kg water without knowing the exact protein concentration in g/l, one may choose to apply the value of w calculated in 7.9.1, dividing therefore by 0.94. To convert mmol to mosmol, the answer needs then to be multiplied by an empirical factor called an 'osmotic coefficient'. This depends on the nature and concentrations of the solutes, but, for a pure solution of NaCl, 150 mmol/kg water, the coefficient is 0.93. This value is about right for plasma.

7.10.2 What is the combined correction factor, to one decimal place?

Thus, for typical blood plasma, the concentration in mosmol/kg water happens to be nearly equal to the sum of all the solute concentrations in mmol/l.

7.11 Gradients of sodium across cell membranes

Sodium is at a much lower concentration inside a typical cell than outside. Because of this gradient, the permeability of the cell membrane to sodium, and the negative membrane potential, sodium tends to leak in, but its concentration is kept low inside by the continual baling action of the sodium pump (Na,K-ATPase). Typically, three sodium ions are transported outwards, and two potassium ions transported inwards, for each molecule of ATP that is hydrolysed to ADP.

There is an upper limit to the electrochemical potential gradient for sodium that the pump can maintain and this is determined by the amount of energy that is available from each mole of ATP. Potassium is also relevant, since this ion is transported by the sodium pump too. However, the argument and calculations are easier if we disregard this ion (which we do only for the moment) and the conclusions are not greatly affected.

The electrochemical potential difference for sodium consists of two terms. One is the membrane potential, V_m, and the other is the equilibrium potential for sodium, E_{Na}. Thus:

$$\text{electrochemical potential difference} = E_{Na} - V_m. \tag{7.12}$$

E_{Na} is related by the Nernst equation to the extracellular and intracellular concentrations of sodium (per kg water), these being $[Na]_e$ and $[Na]_i$ respectively:

$$E_{Na}\,(V) = (RT/F)\,\ln([Na]_e/[Na]_i). \tag{7.13}$$

R is the gas constant, F is the Faraday constant and T is the absolute temperature. For a body temperature of 37 °C the equation may be rewritten more conveniently (using values given in Appendix A) as:

$$E_{Na}\,(mV) = 61.5\,\log([Na]_e/[Na]_i). \tag{7.14}$$

This has appeared before, as part of equation 7.3. As a realistic example, suppose that $[Na]_e$ is 150 mmol/kg water and $[Na]_i$ is 15 mmol/kg water. Then E_{Na} is $61.5 \log 10 = 61.5$ mV.

To return to the electrochemical potential difference, this is also related to the thermodynamic work, W, needed to transport 1 mol of sodium across the membrane, as follows:

$$E_{Na} - V_m\,(V) = W/F. \tag{7.15}$$

If it is assumed now that the sodium pump acts to maintain the maximum electrochemical gradient, then we can relate the latter to the amount of energy available from the hydrolysis of ATP. Assuming that one ATP molecule powers the transport of three sodium ions, W is equal to one-third of $-\Delta G_{ATP}$, the free energy change in the hydrolysis of ATP. Hence, equation 7.15 may be rewritten as:

$$E_{Na} - V_m = -\Delta G_{ATP}/3F. \tag{7.16}$$

The value of $-\Delta G_{ATP}$ depends on the concentrations of the various reactants in the cytoplasm as well as on other physical conditions such as temperature. Although it therefore varies from cell to cell, it may be taken here as about 10–13 kcal/mol. The maximum electrochemical potential difference, $(E_{Na} - V_m)$, thus works out, in volts, at 10–13 kcal/mol divided by $3F$. F is 23.1 kcal/volt equiv. In the case of sodium, 1 equiv = 1 mol.

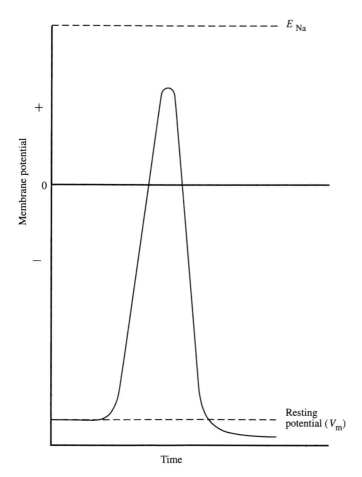

Fig. 7.2. Diagram of a sodium action potential in a nerve fibre. V_m is the resting membrane potential and E_{Na} is the equilibrium potential for sodium.

7.11.1 What is the likely range for the maximum electrochemical potential difference in millivolts?

Let us now see how close the actual electrochemical potentials for sodium are to this range, calculating them from ($E_{Na} - V_m$). For a quick and approximate answer, achieved without having to think about logarithms, it may be recalled that the peak positive potential in a sodium action potential (the overshoot) approaches (without actually reaching) the equilibrium potential for sodium (E_{Na}). This is shown in Figure 7.2.

7.11.2 A textbook gives −90 mV as the resting potential for a nerve fibre and +40 mV for the overshoot of the action potential. What estimate does this yield for the electrochemical potential difference for sodium?

This is obviously a rather cavalier approach to the problem, but the answer is not far from the 144–188 mV just calculated (question 7.11.1) for the maximum that might be produced by the action of the sodium pump. Remember that the peak of the action potential does not actually achieve the sodium equilibrium potential.

Now let us approach the matter in the more obvious way, calculating E_{Na} from concentrations of sodium using equation 7.14. To keep the arithmetic easy, let us start by postulating external and internal concentrations of 150 and 15 mmol/kg water respectively, as above. Let the resting membrane potential again be −90 mV.

7.11.3 What is now the estimate for $(E_{Na} - V_m)$?

This time the answer is within the previously calculated range of 144–188 mV. Note, however, that many estimates for intracellular sodium concentration are lower than our merely convenient choice of 15 mmol/kg water. Note also that chemical estimates of total sodium (e.g. by flame photometry) may include some that is not actually free in the cytoplasm, while $[Na]_i$ in equations 7.13 and 7.14 refers to free sodium.

7.11.4 What would be the new estimate of $(E_{Na} - V_m)$ if the internal concentration were halved?

Is this new estimate also within the range of 144–188 mV that was calculated (in question 7.11.1) as the maximum achievable by the sodium pump? One should hardly take these calculations as implying that the electrochemical potential difference for sodium across all cell membranes is exactly that which can be sustained by sodium pumps transporting three sodium ions per molecule of ATP hydrolysed, but the correspondence is close. In any case, the continual influx of sodium ions, accelerated during action potentials, would tend to keep the electrochemical difference below the theoretical maximum.

The rest of this section is for those who prefer rigour to simplicity; there are no surprises. So far we have only noted in passing the fact that the sodium pump exchanges sodium for potassium. Equation 7.16 may now be revised accordingly. It is assumed that for every three sodium ions transported there are two potassium ions transported, and therefore one net negative charge. The equilibrium potential difference for potassium, E_K, is now required in the calculation. As with sodium, this is given by the Nernst equation, which, for 37 °C, is:

$$E_K = 61.5 \log([K]_e / [K]_i). \tag{7.17}$$

Thus, for example, for $[K]_e = 4.5$ and $[K]_i = 150$, E_K is $-94\,\text{mV}$. The new version of equation 7.16 is as follows:

$$-\Delta G_{ATP}/3F = E_{Na} - \frac{2}{3}E_K - \frac{1}{3}V_m. \tag{7.18}$$

7.11.5 What is the value of the right-hand side of equation 7.18 if V_m is -90 mV and the ionic concentrations, in mmol/kg water, are as before: $[Na]_e = 150$; $[Na]_i = 15$; $[K]_e = 4.5$; $[K]_i = 150$?

Taking the transport of potassium into account thus leads to much the same quantitative conclusion as before (see especially the answer to question 7.11.3, where the internal sodium concentration is the same). This is because V_m is often close to E_K. However, E_K is in fact typically more negative than V_m, as above, so that $(E_{Na} - \frac{2}{3}E_K - \frac{1}{3}V_m)$ is somewhat greater than $(E_{Na} - V_m)$.

7.12 Membrane potentials – simplifying the Goldman equation

Teachers and textbook writers disagree on whether the Goldman equation (or 'constant field equation') is necessary for a basic understanding of membrane potentials. The main aim here is familiarization and simplification. Here is one form of the equation, with V_m being the membrane potential in mV and '61.5' being a constant appropriate at 37 °C (as in equation 7.14):

$$V_m = 61.5 \log \frac{[K]_e + \alpha[Na]_e}{[K]_i + \alpha[Na]_i}. \tag{7.19}$$

The equation may be derived in more than one way, with the meaning of α varying accordingly. This is explained in the next two paragraphs, with minimum mathematical detail.

The Goldman equation is more usually given in a form much like this:

$$V_m = \frac{RT}{F} \ln \frac{P_K[K]_e + P_{Na}[Na]_e + P_{Cl}[Cl]_i}{P_K[K]_i + P_{Na}[Na]_i + P_{Cl}[Cl]_e}$$

$$= 61.5 \log \frac{P_K[K]_e + P_{Na}[Na]_e + P_{Cl}[Cl]_i}{P_K[K]_i + P_{Na}[Na]_i + P_{Cl}[Cl]_e}. \tag{7.20}$$

P_K, P_{Na} and P_{Cl} are membrane permeability coefficients for potassium, sodium and chloride. The equation is valid when there is no net current flowing through the membrane. (Additional terms can be included to allow for the diffusion of other ions through the membrane, but these are not usually significant.) The terms for chloride are commonly omitted, making the equation more like that in the first paragraph (equation 7.19), and this is appropriate when chloride ions are close to equilibrium across the membrane. This they often are, and then, at constant V_m, they have no tendency to carry net current inwards or outwards. Omitting the chloride terms and dividing top and bottom by P_K, we have:

$$V_m = 61.5 \log \frac{[K]_e + P_{Na}/P_K[Na]_e}{[K]_i + P_{Na}/P_K[Na]_i}. \tag{7.21}$$

With α standing for the permeability ratio P_{Na}/P_K, equation 7.21 is equivalent to equation 7.19 above.

Equation 7.21 takes no account of the net ionic current through the cell membrane due to the sodium-potassium pump. This net ionic current exists because only two potassium ions are carried inwards for every three sodium ions transported outwards. There is, however, an alternative derivation of equation 7.19 that does involve this current, although α then has a different meaning. The derivation is mathematically similar to that of the usual Goldman equation and is not detailed here. It starts from the assumption that the net passive diffusional fluxes of potassium and sodium are each equal and opposite to their net non-diffusional fluxes. A steady state is therefore assumed. The ratio of net potassium efflux to net sodium influx, each by passive diffusion, is determined by V_m, P_{Na}/P_K and the ionic concentrations. If this flux ratio is f, then equation 7.19 applies, with α equal to fP_{Na}/P_K. The ratio f applies also to the non-diffusional fluxes and would be 2/3 if these were entirely due to Na, K-ATPase. As to other mechanisms, Na-H counter-transport and Na-K–2Cl cotransport would both raise the value of α. (Alternatively, these two mechanisms may be accommodated in separate terms.)

There are two important points to note about equation 7.19. First, when α is very low (i.e. the membrane is much more permeable to potassium than to sodium), equation 7.19 approximates to the Nernst equation for potassium (equation 7.17) and V_m therefore approaches the equilibrium potential for potassium, E_K. Second, when α is very high (i.e. the membrane is much more permeable to sodium than to potassium, as during the upstroke of the action potential), the equation approximates to the Nernst equation for sodium (equation 7.14) and V_m approaches E_{Na} (as in Section 7.11).

Common textbook values for α, (i.e. P_{Na}/P_K), for mammalian nerve and muscle, are 0.01 or 0.05. For some non-excitable cells, the ratio is much higher, even 0.3–0.5 (with V_m sometimes less than -20 mV). Just so long as one does not deal with particular cells and circumstances, it matters little, because of that variability, exactly what α stands for. For the sake of easy arithmetic, let us now seek further simplification.

7.12.1 Suppose that $[K]_i = 150$ mmol/kg water and $[Na]_i = 15$ mmol/kg water. In relation to the bottom line in equation 7.19, what is $\alpha[Na]_i$ as a percentage of $[K]_i$ when α is (a) 0.05 and (b) 0.5?

It is thus evident that equation 7.19 may be approximated for ordinary resting conditions as:

$$V_m = 61.5 \log\frac{[K]_e + \alpha[Na]_e}{[K]_i}$$

$$= -61.5 \log\frac{[K]_i}{[K]_e + \alpha[Na]_e}. \tag{7.22}$$

This is only valid when α and $[Na]_i$ are not too high, but, because of the logarithmic relationship, the effect on V_m of omitting the term $\alpha[Na]_i$ is further reduced.

For practice calculation, consider a cell for which $[K]_e$, $[Na]_e$ and $[K]_i$ are respectively 4.5, 150 and 150 mmol/kg water. Since graphs showing the dependence of V_m on $[K]_e$ are commonplace, let us vary α instead of $[K]_e$.

(Variations in P_{Na}/P_K, and hence α, may be produced locally by transmitters such as acetylcholine.)

....

7.12.2 What would V_m be for this cell if α were (a) zero, (b) 0.01 and (c) 0.07? (log 3 = 0.477, i.e. about 0.5. log 4 = 2 log 2.)

....

8 Acid–base balance

The most important quantitative relationships in acid–base physiology relate to (1) the definition of pH, (2) the Henderson–Hasselbalch equation (which may also include the relationship between tension and concentration of dissolved carbon dioxide), and (3) the principle of electroneutrality which has already been treated in Chapter 7. It is assumed that these are all to some extent familiar, but the essentials of (1) and (2), and of the subject of buffering, are re-stated here in the knowledge that acid–base balance is a blind spot for many physiologists. As usual, the hope has been to put a new slant on the subject and at the same time provide opportunites for the exercise of familiar concepts and quantities.

The original meaning of 'acid–base balance' is hard to trace, but the term is particularly apt nowadays, since it neatly encompasses three distinct aspects of the topic. The pH of a fluid matters because it governs the ionization of enzymes and other molecules within, or bathed by, the fluid, that is to say the balance of the acid and base forms of imidazole and amino groups especially (i.e. $-NH^+$ and $-N$ in the reaction $-NH^+ = -N + H^+$). The pH is controlled in turn by the balance of a particular acid (i.e. carbonic acid, usually treated in terms of carbon dioxide) and its conjugate base (bicarbonate). The principle of electroneutrality has to do with the balance of 'acids' and 'bases' in an old-fashioned sense, according to which cations (e.g. Na^+ – and even NH_4^+, now seen as an acid!) are often described as bases in the old physiological literature and anions (less often so-described) are 'acids'.

Going beyond what is mainly chemistry (Sections 8.1, 8.2, and 8.5–8.7), there are calculations on aspects of cell pH and bicarbonate (Sections 8.3 and 8.4), the roles of cells and bone mineral in whole-body acid–base balance (Sections 8.8 and 8.9), and the postprandial alkaline tide (Section 8.10).

8.1 pH and hydrogen ion activity

S.P.L. Sørensen introduced the pH notation in 1909, with pH (originally written P_H) standing for $-\log[H]$, [H] being the concentration of hydrogen ions in mol/l. Thus, pH 3 corresponds to 10^{-3} mol/l (1 mmol/l). Most people find this definition adequate for their purposes and for the moment we stay with it.

It is of course more natural, even if the numbers are sometimes cumbersome, to think in terms of concentrations rather than of negative logarithms. Physiologists who think about [H] in preference to pH may know the answer to the next question already. The answer, calculated as 10^{-pH}, is used later, but the question is posed partly to provide an opportunity for mental arithmetic using 'log 2 = 0.3' (Appendix B). The method is given in Notes and Answers.

....

8.1.1 What is the approximate concentration (mol/l or nmol/l) of hydrogen ions in plasma at pH 7.4?

....

Sørensen's first definition runs into difficulties with very small volumes of fluid. Recalling that 1 mol of a substance contains 6.0×10^{23} molecules or ions (Avogadro's number), let us take the case of a solution of pH 7, for which [H] would be 10^{-7} mol/l.

....

8.1.2 If a fluid contains hydrogen ions at a concentration of 10^{-7} mol/l, how many should there be in one cubic micrometre? (1 $\mu m^3 = 10^{-15}$ l.)

....

This answer may be appreciated better in relation to the volumes of cell organelles such as mitochondria. Mitochondria are typically about 0.2–1 μm across and 2–8 μm in length.

....

8.1.3 What is the volume of a cylindrical organelle of diameter 0.2 μm and length 4 μm? (For ease of calculation, take π as 3, as in the Old Testament.)

8.1.4 From 8.1.2 and 8.1.3, approximately how many hydrogen ions would such a volume of fluid contain if the pH were 7.0?

....

This volume is small for a mitochondrion, but not extreme. Moreover, there is within the whole mitochondrion a much smaller space, that between the inner and outer membranes. Even more significant, the inner, matrix space of the mitochondrion is actually much more alkaline than the cytosol (the pH of which may itself exceed 7.0) and would therefore contain only a small fraction of the number of hydrogen ions calculated above. This internal alkalinity is due to the continual translocation of hydrogen ions outwards across the inner membrane that is required to produce the gradient necessary for ATP synthesis (see Section 8.4). Contrast the paucity of hydrogen ions in the mitochondrion with their frequent toing and froing across the inner membrane!

As a general conclusion, there would seem to be well-defined spaces within cells that contain only fractions of hydrogen ions. One is therefore forced to think of pH as reflecting not actual concentrations, but something more like averages in time. The point is that hydrogen ions do not individually exist free for long (they are in any case hydrated in aqueous solutions), but are continually appearing and disappearing as they dissociate from, and reassociate with, hydroxyl ions and buffer molecules.

We have seen one difficulty in defining pH in terms of hydrogen ion concentration. Another difficulty is, that whereas it is easy to prepare solutions of known concentration of, say, sodium, this is not generally true of hydrogen ions. It is only possible to have direct and exact knowledge of [H] in strong solutions of strong acid. It is therefore necessary to define the pH of a solution operationally by comparison with standard solutions of buffer. The defined pH value of a buffer standard is allotted with the intention that it should approximate to $-\log$ (H), where (H) is something like the activity of hydrogen ions in the solution (or, more loosely, their 'effective concentration'). Thus it is that the pH of 0.1 mol/l HCl is not 1.0, but 1.19 (at 20 °C). (Yet another complication, discussed in Section 1.5, is the need to divide the activity by a constant, taken as 1 mol/l.) As the activity of a single ionic species is not strictly definable in theory, the whole concept of pH can be troublesome. However, a proper definition of pH (leaving an obvious question unanswered) is simply this – 'the reading on a correctly calibrated pH meter'. (This is less unsatisfactory than it may seem to be, because the interpretation of pH in the context of ionization reactions involves comparison with dissociation constants, or pK values, that are themselves determined in relation to the pH scale; the two uncertainties therefore cancel out.) When used below, the term 'activity' in relation to hydrogen ions and the abbreviation (H) both mean 10^{-pH}.

8.2 The CO_2–HCO_3 equilibrium: the Henderson–Hasselbalch equation

The Henderson–Hasselbalch equation is fundamental to the quantitative treatment of acid–base balance. Here it is in its most commonly used form:

$$pH = pK_1' + \log \frac{[HCO_3]}{S\,P_{CO_2}}. \tag{8.1}$$

pK_1' is the equilibrium constant for the following reaction:

$$CO_2 + H_2O = HCO_3^- + H^+. \tag{8.1}$$

The value of pK_1' in human blood plasma at 37°C is about 6.1. S is the solubility coefficient for carbon dioxide and its value under the same conditions is about 0.03 mmol/l per mmHg. As noted in Section 5.2, the product $S\,P_{CO_2}$ gives the concentration of dissolved carbon dioxide in mmol/l. For a normal arterial P_{CO_2} of 40 mmHg, it is 1.2 mmol/l, this being about one-twentieth of the concentration of bicarbonate. It is roughly seven hundred times the concentration of carbonic acid, which is therefore a trivial contributor to 'total carbon dioxide'; books having '$[H_2CO_3]$' in place of '$S\,P_{CO_2}$' in equation 8.1 are wrong.

....

8.2.1 At this P_{CO_2} of 40 mmHg, what is the pH when the concentration of bicarbonate is (a) 12 mmol/l and (b) 24 mmol/l? (log 2 = 0.3.)

....

The higher of the two bicarbonate concentrations is in the normal range for arterial plasma (22–30 mmol/l) – and of course this must be so since both the calculated pH and the P_{CO_2} are normal. Within that normal range of arterial bicarbonate concentrations, 24 mmol/l is a convenient value to remember, because it squares exactly with a P_{CO_2} of 40 mmHg and the stated values of S and pK_1'. However, I sometimes use other, equally representative values here, doing so partly for arithmetical convenience, and partly as a reminder of natural variability.

For the free use of the Henderson–Hasselbalch equation with minimum calculation, it is a great help to appreciate that it has the following differential form:

$$\Delta pH = \Delta \log [HCO_3] - \Delta \log P_{CO_2}, \tag{8.2}$$

where Δ denotes a difference or a change in value. Alternatively, for any two conditions 1 and 2:

$$pH_1 - pH_2 = \log\,[HCO_3]_1 - \log\,[HCO_3]_2 - \log\,(P_{CO_2})_1 + \log\,(P_{CO_2})_2$$

$$= \log\frac{[HCO_3]_1}{[HCO_3]_2} + \log\frac{(P_{CO_2})_2}{(P_{CO_2})_1}. \tag{8.3}$$

Note that $pK_1{'}$ and S have conveniently disappeared.

8.2.2 By how much does the pH change if the bicarbonate concentration is halved? (log 2 = 0.3.)

8.2.3 To one decimal place, by about how much does the pH change if the P_{CO_2} rises from 39 to 83 mmHg? (Again use log 2, for only an approximation is required.)

Log 2 has been used four times so far in this chapter. The value of remembering it is stressed in Appendix B and in acid–base balance we have a context in which it can be especially useful. As just illustrated, one may immediately calculate the effects on pH of doubling or halving any initial values of P_{CO_2} or [HCO₃] – or quickly assess other comparable changes, as in question 8.2.3 and the question at the end of Chapter 1. Log 2 has a specific usefulness in graph-drawing, as will now be discussed.

A particularly useful form of graph for illustrating basic principles is one of plasma bicarbonate concentration against P_{CO_2}. Figure 8.1 employs these as axes and shows one point to represent normal values in arterial plasma. A line is drawn to pass through this point and the origin. This corresponds to a constant ratio of bicarbonate concentration to P_{CO_2} and hence to a constant pH (equation 8.1). Whatever the values of $pK_1{'}$ and S, one knows that this pH has to be the normal 7.4, since the two other variables are normal. Two other lines may be added (Figure 8.2) corresponding to pH 7.7 (i.e. 7.4 + log 2) and pH 7.1 (i.e. 7.4 − log 2). The first is obtained, as illustrated in the figure, by plotting a point either for normal P_{CO_2} and twice-normal bicarbonate concentration (A), or for half-normal P_{CO_2} and normal bicarbonate concentration (B). The line for pH 7.1 is obtained in a similar manner. How this 'scaffolding' may be used is illustrated in Figure 8.3, which shows changes in plasma bicarbonate in response to metabolic acidosis and its respiratory compensation. In practice, one may use, and

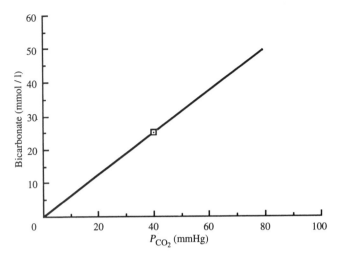

Fig. 8.1. Illustrating the combination of axes recommended for graphing acid–base changes. The single plotted point is for normal arterial plasma. The line drawn through that point shows those combinations of P_{CO_2} and bicarbonate concentration that correspond to the normal pH of 7.4.

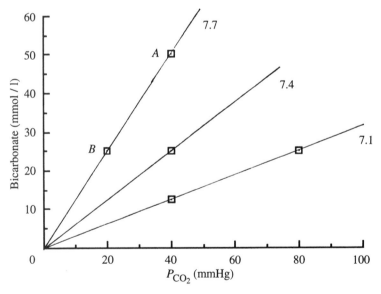

Fig. 8.2. This is the same as Fig. 8.1, except that two more lines of constant pH have been added. The line for pH 7.7 passes through points A and B. A represents a twice-normal bicarbonate concentration and normal P_{CO_2}. B represents a normal bicarbonate concentration and half-normal P_{CO_2}. The line for pH 7.1 is obtained in a similar way.

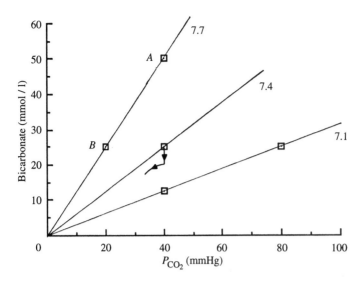

Fig. 8.3. Changes in plasma bicarbonate in response to metabolic acidosis. This is to illustrate the use of the 'scaffolding' of Fig. 8.2, applying it to the particular example of metabolic acidosis. Starting from the point representing normal acid–base balance, the vertical arrow represents the decrease in bicarbonate concentration that constitutes the metabolic acidosis itself; the curve to its left shows the progressive changes in P_{CO_2}, bicarbonate and pH that are associated with respiratory compensation.

expand, only that part of the scaffolding that corresponds to realistic values.

In the preparation of Figures 8.1–8.3 the values of S and pK_1' were not required. In other contexts it can help to combine them into a single constant, as $(pK_1' - \log S)$. Then,

$$pH = (pK_1' - \log S) + \log \frac{[HCO_3]}{P_{CO_2}}. \qquad (8.4)$$

8.2.4 From the above values of pK_1' (6.1) and S (0.03 mmol/l per mmHg), what is $(pK_1' - \log S)$ in plasma at 37 °C?

Combining the constants in this way may save a little time in repeated calculations, but that is not the only point to note. Both pK_1' and S vary with temperature and salt concentration, and are not always readily available for particular conditions that one might happen to be interested in. Then the corresponding value of $(pK_1' - \log S)$ may be calculable from a known combination of pH, P_{CO_2} and bicarbonate concentration. This can be useful in dealing with non-mammalian body fluids, but the question that follows refers to human plasma. It is provided for practice, but the answer may be obvious without further calculation, from what has gone before.

8.2.5 For comparison with the answer to 8.2.4, what is $(pK_1' - \log S)$, given that plasma at 37 °C has a pH of 7.4 when P_{CO_2} is 40 mmHg and $[HCO_3]$ is 24 mmol/l?

The Henderson–Hasselbalch equation is commonly used in calculating bicarbonate concentrations from known values of pH and P_{CO_2}. How accurately should the constants (or the combined constant) be known or specified?

8.2.6 If a bicarbonate concentration were calculated from pH and P_{CO_2}, by what percentage would it be in error if the assumed pK_1' (or else the measured pH) were out by 0.01 unit? ($10^{0.01} = 1.023$.)

The relevant equilibria in plasma and other fluids are more complicated than is usually stated. One reflection of this is that pK_1' decreases somewhat with increasing pH. The exact change in pK_1' per unit change in pH depends on the circumstances (and the experimenter, it seems), but pK_1' has typically been found to fall by about 0.05 for a rise in pH of 1 unit (i.e. $(\Delta pK_1'/\Delta pH) = -0.05$).

8.2.7 In the calculation of bicarbonate concentrations from pH and P_{CO_2}, what percentage difference does this effect make over a pH range of 0.2? (For a quick answer, see the previous question!)

8.3 Intracellular pH and bicarbonate

Are hydrogen ions and bicarbonate ions at equilibrium across typical cell membranes, or do they have a net tendency to diffuse passively either inwards or outwards? The answer is central to our understanding of how cell pH is regulated.

The usual internal pH of mammalian cells seems to vary between about 6.8 and 7.4. Moreover, the pH may fall below 6.8 when lactic acid is produced in anaerobic respiration. It is also true, however, that measured values can depend on the measurement technique and this is necessarily so inasmuch as cytoplasm is heterogeneous. Thus, techniques that measure 'average' cell pH, as opposed to cytosolic pH, are likely to yield higher values when there are many mitochondria present because of their high internal pH. As to extracellular pH, there is normally less variation. The usual pH of arterial plasma is 7.40 and that of mixed-venous plasma is 7.36, while interstitial fluid may be slightly more acid than either. In short, the pH inside a cell may be similar to the pH outside, but is usually somewhat lower.

Given that intracellular pH is usually lower than extracellular pH, one's first thought might be that hydrogen ions tend generally to diffuse outwards. However, ionic diffusion is affected by electrical gradients as well as by concentration gradients and the usual internal negativity of a cell must tend to draw hydrogen ions inwards. To find the net effect of these two opposing forces we may apply the Nernst equation (equation 1.12), as was done previously for other ions (equations 7.3, 7.13, 7.14, 7.17). Accordingly, the equilibrium potential for hydrogen ions, E_H, is given, in mV at 37 °C, by the following equation:

$$E_H = 61.5 \log\{(H)_e / (H)_i\}. \tag{8.5}$$

Note the use of activity here, i.e. (H), rather than concentration as when the equation was presented in Section 1.5. This is actually more proper. Since (H) is defined here as 10^{-pH} (Section 8.1), implying that $pH = -\log(H)$, equation 8.5 may be rewritten in the following form that obviates the need to work with logarithms:

$$E_H = 61.5\{pH_i - pH_e\}. \tag{8.6}$$

For purposes of calculation, here are some definite figures to work with:

$$pH_e = 7.4$$
$$pH_i = 7.0$$
$$\text{membrane potential, } V_m = -70 \text{ mV.}$$

8.3.1 What is the equilibrium potential for hydrogen ions, E_H?

8.3.2 What is the electrochemical potential difference for hydrogen ions across the cell membrane, given by $(E_H - V_m)$?

What is implied by these answers about the direction of net diffusion of hydrogen ions? (It may help to think of the analogy of sodium – Section 7.11.) If you are unsure how to relate signs and directions, try the following approach.

8.3.3 For the same values of pH_e (7.4) and V_m (-70 mV), what would pH_i have to be for hydrogen ions to be in equilibrium across the cell membrane (i.e. with $E_H = V_m$, and V_m substituted for E_H in equation 8.6)?

Since the actual pH is higher than this, it must be that hydrogen ions have a greater tendency to diffuse passively into the cell than out.

We may ask the same questions about bicarbonate ions, but take a short cut to the conclusion. The Henderson–Hasselbalch equation (equation 8.1) relates pH, P_{CO_2} and $[HCO_3]$. Let us assume that SP_{CO_2} and pK_1' are virtually the same inside and outside the cell. Then:

$$pK_1' - \log(SP_{CO_2}) = \log[HCO_3]_i - pH_i$$
$$= \log[HCO_3]_e - pH_e. \tag{8.7}$$

Since $pH = -\log(H)$:

$$[HCO_3]_i \, (H)_i = [HCO_3]_e \, (H)_e,$$

or

$$[HCO_3]_i / [HCO_3]_e = (H)_e / (H)_i. \tag{8.8}$$

Note in passing that equation 8.8 is suggestive of the Donnan equilibrium in its form – cf. equation 7.2 – but clearly it has a different basis. What equation 8.8 actually shows is that bicarbonate and hydrogen ions are quantitatively in an identical state of disequilibrium, except that the directions of net diffusion are opposite – hydrogen ions tending to enter the cell and bicarbonate ions tending to leave. If the basis of this conclusion is unclear, take the logarithm of each side of the equation and multiply by 61.5; the equation then shows that the equilibrium potentials for bicarbonate and hydrogen ions are equal (cf. equation 8.5).

Consistent with this conclusion is the fact that there are mechanisms of membrane transport that act in directions that oppose these tendencies to passive diffusion. These mechanisms vary from one cell type to another, but include Na/H exchange, with sodium entering down its electrochemical gradient and thereby supplying the energy to drive hydrogen ions out of the cell, and also what is effectively a countertransport of HCl and $NaHCO_3$, again driven by sodium entry.

We should not leave this topic without quantifying $[HCO_3]_i$. For this purpose we may use equation 8.7, according to which $(pH_e - pH_i)$ is equal to $(\log[HCO_3]_e - \log[HCO_3]_i)$ and so to $\log([HCO_3]_e / [HCO_3]_i)$.

8.3.4 If the difference between pH_e and pH_i is 0.3 (e.g. 7.4 minus 7.1) and $[HCO_3]_e$ is 26 mmol/kg water, what is $[HCO_3]_i$? ($0.3 = \log 2$.)

This calculation yields merely a representative value for $[HCO_3]_i$. For the same value of $[HCO_3]_e$, and a range of intracellular pH of 6.8–7.4, $[HCO_3]_i$ would be 6.5–26 mmol/kg water. That $[HCO_3]_e$ also varies around the chosen value suggests that $[HCO_3]_i$ might vary even more.

For those who are content at this stage to think in terms of round numbers, the conclusion from the following calculation might be appreciated as an *aide mémoire*. The answer is used later.

8.3.5 If $[HCO_3]_i$ is taken as typically half of $[HCO_3]_e$, and if the mass (or volume) of intracellular water in the whole body is taken as twice the mass of extracellular water, what is the total amount of intracellular bicarbonate in the body divided by the total amount of extracellular bicarbonate?

8.4 Mitochondrial pH

It has been mentioned that the mitochondrial matrix is more alkaline than the nearby cytosol. The pH gradient, and associated membrane potential, are used to drive the synthesis of ATP, and both are set up in the first place using energy released by the passage of electrons along the respiratory chain, this being associated with the membrane itself. Usually the matrix is said to be more alkaline by 'about 1 pH unit' and, for a typical value, it would be unwise to venture any greater precision. The pH is hard to measure *in vivo*, and even *in vitro* it is not easily measured at the correct *in vivo* P_{CO_2}. However, it is easy to calculate a probable upper limit to mitochondrial pH. What is required is an estimate of the local P_{CO_2} and of the maximum bicarbonate concentration, the two of which may then be combined in the Henderson–Hasselbalch equation to yield the pH.

To start with bicarbonate concentration, the higher this is, the higher is the pH. However, the bicarbonate concentration is constrained by the total concentration of solutes, as reflected in the osmolality (Section 7.10), and also by the principle of electroneutrality. Let us start by thinking, simplistically, of the mitochondrial contents as approximating to a solution just of bicarbonate and univalent cation, i.e. a solution of $KHCO_3$ plus a little $NaHCO_3$. The total concentration of the three ions may be taken as having the round-number value of 300 mmol/kg water, the osmolality being similar to that of cellular and extracellular fluids in general. Accordingly, the concentration of bicarbonate is $300/2 = 150$ mmol/kg water. This is the maximum value, given the constraints of osmolality.

As to the P_{CO_2} inside a typical mitochondrion in a normal resting person, this should be close to the intracellular values calculated in Section 5.3, namely between 41 and 51 mmHg. Let us choose a mid-range value of 45 mmHg. The corresponding upper limit to mitochondrial pH may now be obtained from the Henderson–Hasselbalch equation (equation 8.1).

8.4.1 What is the pH if $P_{CO_2} = 45$ mmHg and $[HCO_3] = 150$ mmol/kg water? Take S (the solubility coefficient for carbon dioxide) as 0.03 mmol/kg water per mmHg, and pK_1' as 6.1.

That is the obvious way to make the calculation, but one can easily arrive at a similar answer without using actual values for S and pK_1'. The reasoning is as follows. The P_{CO_2} in a typical mitochondrion is not much higher than in arterial blood. The bicarbonate concentration in arterial plasma is normally about 25 mmol/kg water, so that the concentration in the hypothetical mitochondria (150 mmol/kg water) is higher by a factor of $150/25 = 6$. The term $[HCO_3]/SP_{CO_2}$ in the Henderson–Hasselbalch equation is thus higher by a factor of about 6 also. Accordingly, the maximum mitochondrial pH is higher than the arterial plasma pH of 7.4 by $\log 6 (= 0.78)$. This pH is therefore $(7.4 + 0.78) = 8.18$.

To achieve an accuracy of one decimal place in the calculated pH, the exact choice of P_{CO_2} and osmolality are not very critical. This is because pH does not vary directly with P_{CO_2} and $[HCO_3]$ but with their logarithms. More important is the neglect of other solutes, for their presence could lower [HCO$_3$] substantially. At this point it may therefore be useful to recall the effect on pH of halving [HCO$_3$] at constant P_{CO_2} (question 8.2.2).

8.4.2 By how much is the pH lowered if [HCO$_3$] is halved to 75 mmol/kg water?

The solutes that are probably most significant in this context are anions, notably MgATP^{2-} and other phosphates that are essential for mitochondrial function. Chloride must be present too (but very little because of the membrane potentials of both cells and mitochondria), and a small amount of carbonate (Section 8.6). There must also be some net negative charge on proteins. The concentration of bicarbonate must fall as the concentration of this other material increases, but it is hard to assess how much of the latter is typically present. The best we can conclude, therefore, is that the theoretical maximum pH in the mitochondrial matrix is likely to be 'about 8'. Whether such a pH is achieved *in vivo* is a separate question, but similar values have been measured *in vitro*. For example, Chacon *et al.* (1994) obtained estimates of 7.8–8.2 in cultured myocytes from rabbit heart.

In relation to mammalian cell physiology, measured values may be of more ultimate interest than are theoretical maxima. However, the constraints of osmolality and P_{CO_2} both vary markedly within the animal kingdom and this raises questions for the comparative physiologist. Might mitochondrial pH sometimes be much higher in, say, marine invertebrates of high osmolality and low P_{CO_2}? Could a requirement for a particular optimum range of mitochondrial pH have constrained the evolution of osmolality and P_{CO_2} in any of those animals that have very dilute body fluids?

Calculations more usually presented on mitochondrial pH are of a very different kind and have to do with the mechanism of proton transport and, more precisely, the electrochemical proton gradient across the inner mitochondrial membrane. The situation is somewhat analogous to that of sodium gradients across outer cell membranes and their relationship to membrane potential and free energy changes in the reactions powering ion transport, i.e. the hydrolysis of ATP (Section 7.11). Accordingly, equations 7.12–7.14 may be rewritten in terms of hydrogen ion concentrations and the membrane potential across the inner mitochondrial membrane. To a total electrochemical potential difference of perhaps 220 mV, a pH difference of one unit would contribute about 60 mV. There is nothing in such an analysis to rule out a bigger pH difference.

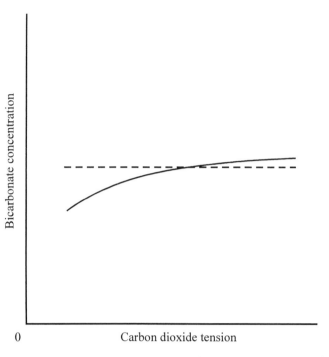

Fig. 8.4. Relationships between bicarbonate concentration and carbon dioxide tension in two solutions. One, like blood plasma, contains non-bicarbonate buffers such as proteins (solid line). The other contains no non-bicarbonate buffer (broken line).

8.5 Why bicarbonate concentration does *not* vary with P_{CO_2} in simple solutions lacking non-bicarbonate buffers

When the carbon dioxide tension rises in a solution containing bicarbonate and some non-bicarbonate buffer (e.g. protein), the concentration of bicarbonate rises too (Figure 8.4). The mechanism is easily understood (see below), but what puzzles some people is the fact that bicarbonate varies hardly at all with carbon dioxide tension (over the physiological range) when the solution contains no non-bicarbonate buffer. The aim here is to clarify that point.

Carbon dioxide reacts with water to produce bicarbonate and hydrogen ions:

$$CO_2 + H_2O = H_2CO_3 = H^+ + HCO_3^-. \qquad [8.2]$$

When there is non-bicarbonate buffer present (represented below as B), the hydrogen ions are mostly buffered as follows:

$$H^+ + B = BH^+.$$ [8.3]

Overall, the reaction is

$$CO_2 + H_2O + B = HCO_3^- + BH^+.$$ [8.4]

It is clear from this that bicarbonate generation is linked virtually one-to-one with non-bicarbonate buffering. Therefore, it should not happen significantly in the absence of non-bicarbonate buffer. And yet the doubt may remain; after all a solution does become more acid when the carbon dioxide tension rises and this implies the production of hydrogen ions in reaction 8.2.

Consider a solution containing sodium, chloride and bicarbonate ions, but no non-bicarbonate buffer. From the principle of electroneutrality (Section 7.6):

$$[Na] + [H] = [Cl] + [HCO_3] + [OH].$$ (8.9)

Two of the terms in this are constant for a given solution, i.e. they cannot vary with carbon dioxide tension. These are [Na] and [Cl]. Thus:

$$[HCO_3] + [OH] - [H] = \text{constant}.$$ (8.10)

Consider now a solution that contains these ions at the following concentrations (chosen to be about right for arterial plasma): bicarbonate 24.00000 mmol/l, hydroxyl ions 0.00010 mmol/l, hydrogen ions 0.00004 mmol/l. The carbon dioxide tension is 40 mmHg. There is no non-bicarbonate buffer present. Next suppose that the carbon dioxide tension is raised sufficiently to double the concentration of hydrogen ions (i.e. [H] becomes 0.00008 mmol/l). Since, at constant temperature, the product ([H] × [OH]) is constant, the concentration of hydroxyl ions is halved. Thus, [OH] becomes 0.00005 mmol/l. (More strictly, it is the product of the activities, (H) × (OH), that is constant.)

....

8.5.1 According to equation 8.10, what is the new concentration of bicarbonate? Is it measurably different from 24.00000 mmol/l?

8.5.2 What is the final carbon dioxide tension?

....

To some this may seem pernickety book-keeping; a man with a million pounds may spend a penny to significant effect, yet clearly leave undiminished his status of millionaire.

One anion that must be present with the bicarbonate has been ignored so far, namely the carbonate ion. It has been ignored mainly for simplicity, but also because it is not generally what the aforesaid puzzled people have in mind, and there is very little present at pH 7.4. We look at carbonate next.

8.6 Carbonate ions in body fluids

The carbonate ion rarely features in elementary accounts of mammalian acid–base balance. Following on from the previous section, it is discussed here mainly in relation to the dependence of bicarbonate concentration on carbon dioxide tension. However, carbonate is also of interest to physiologists because of its presence in bone and in the shells of birds' eggs and of molluscs and crabs (where it occurs as calcium carbonate). To those experimenters that prepare physiological salines, carbonate can be a nuisance if its concentration is allowed to rise so high that it precipitates with calcium.

Carbonate forms from bicarbonate thus:

$$HCO_3^- = CO_3^{2-} + H^+. \tag{8.5}$$

At pH 7.4 the concentration of carbonate in blood plasma is about 0.1 mmol/l. That should also be true of the plasma-like solution described in the previous section, though, for simplicity, this was ignored. With the postulated doubling of P_{CO_2} and consequent lowering of pH to 7.1, the concentration of carbonate would be lowered, but, as with hydrogen and hydroxyl ions, its concentration starts low enough that only tiny amounts of extra bicarbonate could be formed.

With substantially falling P_{CO_2}, hence rising pH, it is a different matter; then, at tensions very much below normal, the concentration of bicarbonate declines markedly. As bicarbonate ions dissociate (i.e. reaction 8.5 proceeds to the right), the released hydrogen ions tend to decompose other bicarbonate ions in the following overall reaction:

$$2HCO_3^- = CO_3^{2-} + CO_2 + H_2O. \tag{8.6}$$

The relationship between carbonate and bicarbonate is given by the following equation, in which pK_2' is the dissociation constant for reaction

8.5 (being, in other words, 'the second dissociation constant of carbonic acid', hence the subscript).

$$\text{Log}\frac{[CO_3]}{[HCO_3]} = pH - pK_2'. \tag{8.11}$$

8.6.1 If pK_2' is 9.8, what is the ratio of carbonate to bicarbonate at (a) pH 7.8, (b) pH 8.8, (c) pH 9.8?

Consider again a solution containing 24 mmol/l bicarbonate and no non-bicarbonate buffer. Its P_{CO_2} is 40 mmHg and its pH is 7.4. Now suppose that carbon dioxide is lost until the pH rises to 7.8. In accordance with the last answer and with reaction 8.6, only about 2% of the bicarbonate is converted to carbonate. Since the pH rises above 7.7, and therefore by more than 0.3 units, the P_{CO_2} must evidently fall somewhat below half of the initial 40 mmHg. (Note yet again the use of log 2 to achieve a rough answer.) The true P_{CO_2} is actually about 16 mmHg.

Plasma does not generally become more alkaline than pH 7.8 in respiratory alkalosis, but let us consider the effects of further loss of carbon dioxide, as when blood is exposed to air. If the pH were to rise to 8.8, then the solution would contain 20.0 mmol/l of bicarbonate and 2.0 mmol/l of carbonate. (The total negative charge, equal to $[1 \times 20.0 + 2 \times 2.0]$ mequiv/l, would be the same as when the bicarbonate concentration was 24 mmol/l and the carbonate concentration was close to zero.) The P_{CO_2}, calculated from the Henderson–Hasselbalch equation, would be only 1.3 mmHg. To check this P_{CO_2}, equation 8.4 may be used with $(pK_1' - \log S)$ taken as 7.62 (the value calculated in question 8.2.5).

8.6.2 As a check, is the pH equal to 8.8 when the P_{CO_2} is 1.3 mmHg and the bicarbonate concentration is 20 mmol/l?

Despite what was said in Section 8.5, bicarbonate concentrations in solutions free of non-bicarbonate buffer do fall substantially when the carbon dioxide tension is reduced far enough. The pH is then well above even pathological values, so that the fall in bicarbonate concentration is of small importance *in vivo*.

8.7 Buffering of lactic acid

Human blood plasma typically contains about 0.7–2.0 mmol/l lactate, but the concentration can rise as a result of anaerobic metabolism, as in exercise, and as a compensatory response to alkalosis. Although lactate in body fluids is often referred to as 'lactic acid', it is actually nearly all ionized. The next question is for those who would like to explore that point before returning to the matter of buffering.

The ratio of lactic acid to lactate is given by the following equation, this being derived by rearrangement of one that defines the equilibrium constant, pK.

$$\text{Log} \frac{[\text{lactic acid}]}{[\text{lactate}]} = pK - pH. \tag{8.12}$$

The pK is about 4.6 in human plasma.

8.7.1 What is the ratio of lactic acid to lactate at (a) pH 6.6 and (b) pH 7.6? (Most human body fluids are within this range of pH.)

Consider now how much the pH of extracellular fluid falls on addition of a given amount of lactic acid. The hydrogen ions are buffered by bicarbonate (with release of carbon dioxide) and to a lesser extent by proteins. The inorganic phosphate of extracellular fluid contributes rather little to buffering, but further hydrogen ions are buffered by erythrocytes, and by other tissues too in the long term.

There is an obvious advantage in considering a simpler situation here, so let us suppose that 5 mmol/l lactic acid is added to a salt solution of pH 7.4 containing 25 mmol/l bicarbonate and no other buffers. Of that bicarbonate, 5 mmol/l is turned to carbon dioxide and water, but the carbon dioxide is then lost, we also postulate, with readjustment of the P_{CO_2} to its original value. The reaction is as follows:

$$CH_3CH(OH)COOH + HCO_3^- = CH_3CH(OH)COO^- + H_2O + CO_2. \quad [8.7]$$

The point of interest now is the fall in pH. Since the P_{CO_2} returns to its original value, the one change that is relevant to calculating the new pH is the fall in bicarbonate concentration from 25 to 20 mmol/l. The Henderson–

Hasselbalch equation could be used as it stands, i.e. equations 8.1 or 8.4, but it is easier to use equation 8.3. With P_{CO_2} constant, the terms involving P_{CO_2} drop out, leaving the following:

$$pH_1 - pH_2 = \log \frac{[HCO_3]_1}{[HCO_3]_2}. \tag{8.13}$$

8.7.2 In the above circumstances, what is the final pH when the bicarbonate concentration is lowered from 25 to 20 mmol/l, when the P_{CO_2} is unchanged, and when the initial pH is 7.4?

This calculation provides an estimate of the fall in pH *in vivo* when P_{CO_2} is held constant through adjustments in ventilation, but P_{CO_2} would actually be adjusted to a lower value that would reduce the acidaemia (respiratory compensation). The acidaemia would also be reduced by the buffering effect of plasma proteins and, more slowly, by buffering within cells (Section 8.8) and bone (Section 8.9). The calculated fall in pH is therefore an overestimate of the fall *in vivo*.

Suppose now that all the carbon dioxide released from bicarbonate by the lactic acid were to stay in solution – with a consequent rise in P_{CO_2}. The concentration of bicarbonate falls again from 25 mmol/l to 20 mmol/l, but this time the concentration of dissolved carbon dioxide rises by 5 mmol/l. Its initial concentration can be taken as 1.2 mmol/l – normal for arterial plasma, like the initial pH and bicarbonate concentration. Thus the concentration of dissolved carbon dioxide rises by a factor of $(5 + 1.2)/1.2$, or about 5 – and with it the partial pressure.

8.7.3 What would the final pH be this time? The Henderson–Hasselbalch equation may be used again, but, for easy mental arithmetic, start with the answer to the previous question and adjust it for the (near-enough) fivefold increase in P_{CO_2} by applying equation 8.3, noting that $\log 5 = (\log 10 - \log 2) = 0.7$.

The difference between the two answers illustrates the special nature of the HCO_3/CO_2 buffer system – the fact that one component can be regulated by adjustments in pulmonary ventilation.

8.8 The role of intracellular buffers in the regulation of extracellular pH

Acid–base disturbances may be respiratory or metabolic and may involve acidosis or alkalosis. For simplicity, we concentrate here on respiratory acidosis. When the mean body P_{CO_2} is increased, the resulting fall in extracellular pH is moderated by a variety of mechanisms that raise the bicarbonate concentration in the extracellular fluid. Important amongst these is buffering by plasma proteins, erythrocytes, muscle and other cells, and bone. In addition, bicarbonate is generated by the kidneys ('renal compensation') and, to a small extent, by reduction in the concentration of organic anions, mainly lactate. The role of the extrarenal mechanisms has been studied in nephrectomized mammals.

The rise in extracellular bicarbonate concentration in response to a given degree of respiratory acidosis varies, but a doubling of P_{CO_2} may lead – over an hour or so, and by extrarenal mechanisms – to a rise of about 3 mmol/l. The purpose of the next calculation is to quantify the usefulness of this in moderating the initial fall in pH. Remember that the fall in pH due directly to a doubling of P_{CO_2} is 0.3 units (Section 8.2).

....

8.8.1 For comparison with this fall in pH of 0.3 unit, what is the effect on pH of raising the bicarbonate concentration from 25 to 28 mmol/l?

....

One might suspect that this unimpressive effect is largely due to erythrocytes. If P_{CO_2} is raised from 40 to 80 mmHg in a sample of oxygenated blood *in vitro*, the plasma bicarbonate concentration rises by about 7 mmol/l (depending on the haematocrit and other factors). This is much more than the rise *in vivo*, but it must be remembered that the bicarbonate released from erythrocytes *in vivo* is shared with the interstitial fluid as well as with the plasma.

....

8.8.2 *In vitro*, the concentration of bicarbonate in the plasma of a particular blood sample rises by 7 mmol/l when the P_{CO_2} is doubled. What would the corresponding rise be if the same amount of bicarbonate left each erythrocyte, but (much as *in vivo*) became distributed in 15 l of extracellular fluid instead of in 3 l of plasma?

....

If the actual rise *in vivo* is 3 mmol/l (see above), it would thus seem that erythrocytes are not responsible for all of it. This is true. However, it is not as easy as it might seem to calculate (by difference) the contribution of the erythrocytes, for the situation *in vivo* is complicated. One reason is that the blood is partly oxygenated and partly deoxygenated, and this affects the buffer properties of the haemoglobin. Also, the quantity of bicarbonate released from the erythrocytes is influenced by the final concentration in the plasma and this is determined not only by the relative volumes as in 8.8.2, but also by the amounts of bicarbonate entering the extracellular fluid from elsewhere (see Notes and Answers). A better estimate of the contribution of erythrocytes to buffering requires more information, and more complicated calculations than are appropriate here.

The buffering of extracellular fluid by bone mineral and by cells other than erythrocytes ('tissue buffering') receives little or no mention in many accounts of acid–base physiology. Our calculations so far suggest why; the effect would seem to be small. There is actually much more to the phenomenon and, to see why, we must first consider the regulation of cell pH.

Much of the work on tissue buffering of extracellular fluid was carried out decades ago. Nowadays there is more interest in the study of pH regulation in single cells. To take again the case of raised P_{CO_2}, this initially makes the cytoplasm more acid. The cell must then increase its content of bicarbonate if it is to recover its original pH. It may do this partly by intracellular buffering and partly by transport of ions between cytoplasm and extracellular fluid – bicarbonate inward or hydrogen ions outward. (These two processes are equivalent in effect, for both of them lead to an increase in the bicarbonate content of the cell and a loss of bicarbonate from the extracellular fluid.) A very small amount of bicarbonate may also be generated by metabolism of organic anions such as lactate.

So, to reiterate, the cell with a raised P_{CO_2} achieves better pH regulation if it increases its bicarbonate content rather than losing bicarbonate to the extracellular fluid. There is thus a potential conflict between the regulation of intracellular pH and the regulation of extracellular pH.

To emphasize this conflict between intracellular and extracellular pH regulation, let us consider a much simplified model of the body. It has just two fluid compartments – extracellular and intracellular. There are no kidneys, bone, or erythrocytes and no non-bicarbonate buffers (e.g. proteins) in the extracellular fluid. The P_{CO_2} is the same in each compartment.

The concentrations of organic anions do not change. There is no osmotic redistribution of water following the movement of ions between body compartments.

Let us make the extreme assumption that all of the cells are able to regulate their pH perfectly by membrane transport of hydrogen or bicarbonate ions. This has the convenient effect of making buffering irrelevant, as will now be explained. When P_{CO_2} is first raised, there would indeed be production of bicarbonate within the cells as a result of buffering (by proteins, dipeptides, phosphates, etc.) as in reaction 8.4, but as the intracellular pH returns to normal, the ionization of the intracellular buffers, which is pH dependent, also returns to normal. It follows that the ultimate contribution of buffering is nil, and this means that the total amount of bicarbonate in the two body compartments, taken together, stays constant. As to the amounts of bicarbonate originally present in each of the two individual compartments, it suffices here simply to take them as equal. (For the basis of this assumption, see the answer to question 8.3.5.)

We may now explore the effect of a rise in P_{CO_2} on extracellular bicarbonate and pH. Although the calculations relate to a much-simplified model of the body, the qualitative conclusions will be clear and important.

....

8.8.3 For cell pH not to change, by what factor would the intracellular bicarbonate have to rise if P_{CO_2} were to double? (See equation 8.1.)

8.8.4 Given that the amounts of bicarbonate in the two compartments start equal, and that their total stays constant, how much bicarbonate would be left in the extracellular compartment after the doubling of P_{CO_2}?

....

Worded in a way that is more appropriate to a real body, the implications of the calculation are as follows: if most of the cells are able to maintain a normal or nearly normal pH in the face of a raised P_{CO_2}, then the extracellular bicarbonate concentration and pH should fall substantially. This is not what happens in mammals, nephrectomized or not (though the effect has been observed in a species of salamander).

Of the many simplifying assumptions in the model, the one most significant in accounting for its unrealistic behaviour is almost certainly the assumption that the cells maintain perfect pH regulation. (The non-bicarbonate buffers of extracellular fluid, including plasma proteins, were excluded from the model, but are of little quantitative importance. The

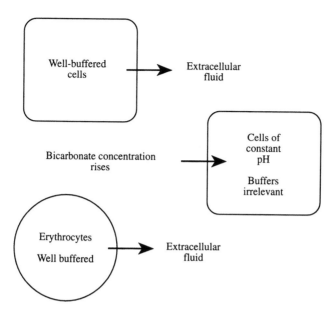

Fig. 8.5. Movements of bicarbonate ions (or of hydrogen ions in the opposite directions) between cells and extracellular fluid in response to raised P_{CO_2} (respiratory acidosis). The diagram represents a model organism in which some cells are able to maintain constant internal pH.

inclusion of erythrocytes would only make the model marginally more realistic.) Further analysis of what occurs in the real body in response to respiratory acidosis is beyond the scope of simple calculation, but what seems to happen is summarized in Figure 8.5. Those cells best able to regulate their internal pH increase their bicarbonate content at the expense of the extracellular fluid. Other cells (including erythrocytes) produce bicarbonate by internal buffering and share some of it with the extracellular fluid. An uncertain amount of bicarbonate is also produced by buffering at the surface of bone mineral (see Section 8.9).

There is thus more movement of bicarbonate about the body than is implied by the small increase in plasma bicarbonate that is observed in the extrarenal response to respiratory acidosis. The redistribution of bicarbonate is all the more important because it is accompanied by movements of potassium and sodium in accordance with the principle of electroneutrality. Thus it is that the concentration of potassium in plasma tends to rise in acidosis. Changes in the distribution of sodium are less obvious, for reasons discussed in Section 7.3.

8.9 The role of bone mineral in acid–base balance

It is evident from the formulae of the various forms of bone mineral given in Section 7.4 that their precipitation and dissolution must affect acid–base balance. Moreover, it is common knowledge that bones may be dissolved in acid (bones of murder victims, bone in a dog's stomach), so it is obvious that dissolution must be accompanied by depletion of acid and by a rise in pH.

The extracellular fluid contains bicarbonate at a concentration of about 25 mmol/l and the alkalinizing effect of the dissolution of bone is to raise that concentration. We may explore the magnitude of that effect in terms of the ratio of bicarbonate generation to calcium release. For all the forms of bone mineral mentioned above that ratio is fairly close to unity. The next paragraph shows how it is derived for the case of $Ca_9(PO_4)_{4.5}(CO_3)_{1.5}(OH)_{1.5}$. This is chosen as the most complicated example, as well as the most abundant; the reader may choose to explore the simpler three unaided.

The following equation represents the process of dissolution into a bicarbonate-containing solution at pH 7.4:

$$Ca_9(PO_4)_{4.5}(CO_3)_{1.5}(OH)_{1.5} + 8.4\ CO_2 + 6.9\ H_2O = 9\ Ca + 3.6\ HPO_4$$
$$+ 0.9\ H_2PO_4 + 9.9\ HCO_3. \qquad [8.8]$$

It may not be immediately obvious where the various coefficients come from, but the starting point is the fact that the ratio $[HPO_4]/[H_2PO_4]$, which is pH dependent, is equal to 4 at pH 7.4: accordingly $3.6/0.9 = 4$ and $(3.6 + 0.9) = 4.5$, the number of phosphates in the formula. (Concentrations of PO_4 in extracellular fluid, and likewise CO_3 and OH, are all small enough to ignore here, and this applies also to the concentration of non-bicarbonate buffer.) Reaction 8.8 shows that 9.9 bicarbonate ions are produced for each 9 calcium ions released. For the purpose of answering the next question, the ratio may be taken as near enough 1.0.

....

8.9.1 Suppose that the total concentration of calcium in the extracellular fluid is raised (very substantially) by 1 mmol/l through the dissolution of the specified bone mineral. What would be the final concentration of bicarbonate if the initial value were 25 mmol/l? (Ignore again the small effects of non-bicarbonate buffers and treat the extracellular fluid as a closed system with respect to ions.)

8.9.2 What would be the change in pH if P_{CO_2} were maintained constant?

....

In interpreting this, note that, with care, plasma pH may be measured with capillary electrodes to a precision of about 0.002. Routine measurements on other solutions with the usual large electrodes do not generally have an accuracy better than 0.02.

These calculations seem to lead to a clear conclusion about the role of bone mineral in buffering – that it is small – but two very important points need to be made in qualification. The first is that bone phosphates may have a significant role in the buffering of acid in the long term, provided that the dissolving calcium and phosphate are continually excreted so that plasma levels do not change. The second point is that bone mineral does not consist just of the above salts. Also present are sodium, potassium and what seems to be bicarbonate. It is likely that these are released in response to acidosis, as a separate and important mechanism of buffering.

8.10 Is there a postprandial alkaline tide?

For each hydrogen ion secreted into the stomach by the parietal cells, a bicarbonate ion is secreted into the blood. The tendency is therefore slightly towards metabolic alkalosis. Many textbooks refer to a consequent 'postprandial alkaline tide', a temporary reduction in the excretion of acid by the kidneys. The idea dates back to Bence-Jones (1845), but few books give any quantitative information on the phenomenon.

Here we try to estimate the possible magnitude of the tide and of the associated change in plasma pH. For this we need to know how much acid is secreted after a typical meal, and how quickly. Amounts vary considerably from person to person and from meal to meal, with maximal rates varying from 1 to 47 mequiv/h (see Notes and Answers). Let us choose for our arithmetic a modest total secretion of 28 mequiv.

To calculate the effect on blood pH accurately, we would need to take into account the buffering that occurs in the plasma and other extracellular fluid, in the erythrocytes and other cells, and in bone. We also need to consider renal compensation and the time courses of all of these. The exercise is clearly too complicated for the back of an envelope, so let us start with an easier calculation that is based on gross oversimplification.

Accordingly, let us suppose that an individual secretes 28 mequiv of gastric acid, that all of this comes from 14 l of extracellular fluid, and that erythrocytes and the rest of the body make no contribution whatsoever to buffering and the adjustment of extracellular pH. (By excluding these other mechanisms we overestimate the effect on pH.) The extracellular fluid initially contains, let us say, 25 mmol/l bicarbonate and no non-bicarbonate buffer, so that the loss of acid means an equivalent gain of bicarbonate. Secretion of gastric juice implies loss of extracellular fluid, but let us ignore that.

8.10.1 In this simplified situation, what is the final concentration of bicarbonate after secretion of the acid?

8.10.2 If there were no respiratory compensation (beyond removal of the excess CO_2 produced by the decomposition of bicarbonate) and the P_{CO_2} were thus to stay constant, by how much would the pH of the extracellular fluid rise?

The rise is small, but detectable. It should be less in a real body because (a) P_{CO_2} would tend to rise in compensation, and (b) other acid–base mechanisms would come into play to lower the extracellular bicarbonate concentration. Alternatively, we could have chosen a greater quantity of acid secretion.

To consider possible effects on renal excretion, an easily found item of quantitative information is that typical rates of acid excretion, averaged over 24 hours, are about 40–110 mequiv/day, these figures representing the sum of titratable acid and ammonia. The equivalent hourly rates are 1.7–4.6 mequiv/h, but in this shorter term a wider range would be expected, including negative values when there is sufficient bicarbonate excretion. Given these rates, one might well expect the gastric generation of bicarbonate at 1–47 mequiv/h to produce a conspicuous reduction in acid excretion sometimes. Nevertheless, several investigators have sought the alkaline tide without finding it. On the heels of gastric acid secretion comes the pancreatic secretion of bicarbonate and this must surely diminish the tide. Indeed, it seems appropriate that the two processes should cancel out, more or less, both in the gastrointestinal tract and in the extracellular fluid. The generality of the postprandial alkaline tide remains unclear (see Notes and Answers). Note, however, that the background level of acid excretion does itself relate to food; for example, the oxidation of sulphur in ingested protein to sulphate plus hydrogen ions ('sulphuric acid').

9 Nerve and muscle

Here is a field of physiology that is full of quantification and mathematics, but opportunities for applying just shop-keeper's arithmetic are fewer than might be expected. In Sections 9.1 and 9.2 we look at conduction in nerve and cardiac Purkinje fibres. Section 9.3 takes us a tiny step in the direction of integrated neuromuscular activity, but also has the humbler aim of supplying a context in which to think about time scales and the speeds of neuromuscular events. Section 9.4 relates muscle mass to the activities of chinning the bar and jumping, in a simple, rough-and-ready way that avoids calculating forces in complicated lever systems. The approach leads naturally to a consideration of creatine phosphate usage in muscle contraction (Section 9.5). Finally, Section 9.6 integrates information on sarcomere dimensions, myofilament spacing and calcium concentration in the context of muscle activation. A few related topics are treated earlier in the book: the energetic efficiency of skeletal muscle in Section 3.9, tensions in arteriolar smooth muscle in Section 4.5, and resting membrane potentials and action potentials in Sections 7.6, 7.11 and 7.12.

One may memorize the numerical dimensions of skeletal muscle fibres, yet not perceive that some of these fibres have the lengths and widths of suitably chosen human hair. A class of students beginning physics was asked to estimate the height of the Empire State Building that could be seen from the window (1250 ft, or 381 m, without its television aerial). Estimates ranged from 50 ft (15 m) to 1 mile (1.6 km). How much harder it must be to feel comfortable with microscopic dimensions or milliseconds! If a stereocilium in the organ of Corti were scaled up mentally to the size of the Empire State Building, how much would it sway in response to a whisper? Many other such problems hover on the edge of arithmetic and a handful is addressed below.

9.1 Myelinated axons – saltatory conduction

Myelinated axons vary in conduction velocity from a few m/s to 120 m/s. Consider a peripheral nerve fibre with a conduction velocity of 100 m/s and with internode lengths of 1.5 mm, figures appropriate to a motor fibre with a diameter of about 18 μm.

9.1.1 How long does it take an action potential to spread from one node of Ranvier to the next?

This answer may be compared with the duration of a single action potential at a node. Figures in typical elementary textbooks depict the positive phase of a nerve action potential as lasting about 0.5–1 ms, with the peak occurring at 0.2–0.6 ms. These durations are all more than ten times the answer to question 9.1.1. What does this mean? The point here is that some of those books give the clear impression (which many others do not contradict) that saltatory conduction in myelinated nerve fibres involves a 'jump' from one node to the next, with a pause for regeneration at each before activation of the next node. Perhaps the word 'saltatory' (= leaping) seems to imply this too, though that was never the intention. What the calculation implies, and what has long been known, is that the action potential constitutes a wave of activity that spans many nodes simultaneously.

The above myelinated nerve fibre has an internode length of 1.5 mm and a diameter of about 18 μm. Most myelinated fibres are narrower than this, but have about the same ratio of internode length to diameter. It is important to visualize these relative dimensions, and casual microscopy may not help much, for, examined with a ×40 microscope objective, many Schwann cells are far longer than the field diameter. Nor are all textbook diagrams much help in this regard if they are like the typical examples shown in Figure 9.1.

9.1.2 If the above myelinated fibre is drawn with correct proportions, but with a diameter of 18 mm (i.e. ×1000), what length of paper would be required to show two nodes of Ranvier?

Conducted at 100 m/s, an action potential in the above nerve fibre would travel 1 m in 10 ms. In the case of a very narrow myelinated fibre with a diameter of, say, 2 μm, the conduction velocity could be about 10 m/s.

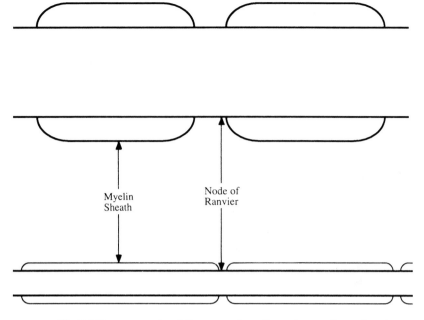

Fig. 9.1. Two conventional diagrams of myelinated nerve fibres, showing myelin sheaths and nodes of Ranvier.

9.1.3 What difference, in ms, would this roughly tenfold reduction in diameter make to the time required for a motor response involving motor fibres (e.g. to muscles in the foot) that are 1 m in length?

9.2 Non-myelinated fibres

C fibres are non-myelinated. The fastest conduct at about 2.3 m/s and have diameters of about 1.1 μm. While conduction velocity is roughly proportional to fibre diameter in myelinated fibres, it is proportional to the square root of fibre diameter in non-myelinated fibres. From this information, with the diameter in μm and velocity in m/s:

$$\text{velocity} = (2.3/\sqrt{1.1})\ \sqrt{(\text{diameter})} = 2.2\sqrt{(\text{diameter})},$$

or

$$\text{diameter} = (\text{velocity})^2/4.8. \tag{9.1}$$

9.2.1 The widest myelinated fibres conduct at about 120 m/s. Assuming the correctness of equation 9.1 for any diameter, how wide would a non-myelinated fibre have to be to have the same conduction velocity?

Extrapolating like this is almost always unwise, but the general conclusion is safe: without myelination, our motor nerve fibres (and the nerves in which they run) would have to be very thick to conduct as rapidly as they need to.

Purkinje fibres in the heart need to be wide in order to conduct quickly enough for a rapid and properly coordinated heartbeat. With reservations, equation 9.1 may be tried on them as a test of its generality.

9.2.2 A Purkinje fibre of diameter 100 μm may conduct at about 4 m/s. What is its conduction velocity calculated from equation 9.1?

This is hardly excellent agreement, but it is not as bad as might be anticipated, considering the differences between the action potentials in nerve fibres and Purkinje fibres, and the great extrapolation involved.

9.3 Musical interlude – a feel for time

We all have a feeling for hours, minutes and seconds. The milliseconds of physiology may need more thinking about to become meaningful. A vibration in the vocal cords of 1000 cycles/s (i.e. 1 cycle/ms) gives a musical pitch about one octave above 'middle C'. The note from a 32-foot (9.8-m) organ pipe may throb with a frequency near 16 cycles/s, about as low as most people are able to perceive as a note rather than as separate sounds (which would be spaced at about 60 ms).

A book of transcribed solos of the jazz saxophonist Charlie Parker shows that he improvized on the tune *Bird gets the worm* at a rate of 340 beats per minute (this being the fastest rate in the collection). Mostly he was playing two notes (quavers) per beat, though sometimes three or four. Playing quavers he was thus fingering notes at a rate of 11/s.

9.3.1 At that rate, how many milliseconds were available for each note?

Saxophones and other musical instruments can be played faster, especially when there is no improvization involved and the fingers follow well-practised patterns of movement. Playing with a frequency of, say, 13 notes/s (1 per 77 ms), one even approaches the throb rate of a deep organ pipe. For comparison, note that the syllables of human speech can barely exceed 10/s (though the motor commands needed for speech must number hundreds per second). As there is a similar limitation to mental counting, it is not actually very easy to judge the pace of the fastest music.

Part of the point of the above calculation is that it may give one a better feel for time intervals close to 100 ms (0.1 s), and a context in which to consider the time scales of simpler neural and neuromuscular phenomena. Here are some examples, although it is best to think about the matter in the context of one's own knowledge and interests. In a (long) myelinated axon with a (slow) conduction velocity of 10 m/s, 100 ms is the time it takes an action potential to travel 1 m. The time elapsing between the start of an isometric contraction and the moment of peak tension in various human hand muscles after a single electrical stimulus to the nerve is 55–80 ms, while calf muscles may take twice as long to achieve peak tension. About 20 ms elapses between stimulation of the motor cortex and resulting action potentials in the finger muscles. Eye closure during a blink takes about 50–150 ms and reopening takes about 100–200 ms. Elite sprinters may take four to five strides per second, at 200–250 ms/stride. Humming birds beat their wings at 30–50 times per second, so that a single complete beat may take as little as 20 ms.

To return to the subject of music, even the most relevant of those examples is no explanation for the limits to fast instrumental playing. It is not the case that the fingering of each note is separated in time from the fingering of the next, for the motor activities overlap. This is less obviously the case with a trill, since this typically involves a single finger moving rapidly up and down. Depending on the finger and the person, the rate of movement could be 6–9/s. Since each movement in a trill produces two notes, the time for each note is 56–83 ms.

With regard to Charlie Parker, there is one particular question to be addressed: to what extent was there time for him to hear and react to one note before he played the next? A player can certainly do this in much slower music.

Consider 'reaction time' in the sort of experiment in which a subject presses a button in response to a sensory stimulus. If the latter is a sudden light, then the reaction time could be perhaps 150–250 ms, depending very much on the individual and circumstances. Reactions to sound, more relevant here, are usually about 30–50 ms faster. The reactions of elite sprinters to a starting gun can be as quick as those two ranges suggest. Reactions requiring a decision, such as which of several buttons to press, take substantially longer.

9.3.2 Which is longer, the answer to 9.3.1 or the reaction time to an auditory stimulus?

Jazz musicians are continually responding to the notes that they and the other musicians produce, but in fast playing there is clearly no possibility of useful feedback between successive notes. Rather, the fingers are responding to pre-programmed patterns of neural activity that correspond to groups of notes. How this is achieved, and without any feeling of delay, is hardly something to solve by simple arithmetic.

9.4 Muscular work – chinning the bar, saltatory bushbabies

The mechanics of human movement would seem to offer considerable scope for quantitative treatment involving muscle tensions and the arithmetic of levers. However, a simple limb movement may call on a number of muscles working together, and the relevant measurements may be hard to elicit, from even shelves of anatomy books. Here we side-step these complexities to utilize an idea of seductive simplicity: the work of which a skeletal muscle is capable, in a single contraction *in situ*, is roughly proportional to its mass or volume. The reasoning behind that is as follows.

The forces that skeletal muscles exert are proportional to their cross-sectional areas. The distances that they are able to shorten *in situ* are proportional to their resting lengths (and also dependent on their anatomical location). The work that they can do in a single contraction is given by the mean force multiplied by the shortening distance, and is therefore proportional to the product of cross-sectional area and length, and hence to volume. Muscle volume is proportional to muscle mass.

To develop these ideas in a more detailed, quantitative argument, let us start by considering an idealized muscle with a uniform cross-sectional area throughout its length. Its initial length is L m, its initial cross-sectional area is A m^2, and its volume is therefore LA m^3. Under their normal conditions *in vivo* (i.e. constrained by the skeleton and tendon elasticity) the locomotory skeletal muscles are said to be able to shorten, typically, by up to 25%. (Some put the figure at 30%, but muscles vary in their ability to shorten and it is hard even to define an average.) So let us assume a shortening distance here of $0.25L$ m. Over this limited range the tensions that muscles are able to exert are close to those at the maxima of their length–tension curves. This means that, when contraction is very slow (or isometric), stresses can be about 3 kg-force/cm², or 3×10^5 N/m². The force exerted by our idealized muscle can therefore be taken as $(3 \times 10^5 \text{ N/m}^2) \times (A\text{m}^2) = 3A \times 10^5$ N. The work done in a slow contraction, calculated as force times distance, is $(3A \times 10^5 \text{ N}) \times (0.25L\text{m}) = 0.75AL \times 10^5 \text{ Nm} = 0.75AL \times 10^5$ J. Dividing this by the volume of muscle, LAm³, we get the work done per unit volume, i.e. 0.75×10^5 J/m³. To express this as work per unit mass, we divide it by the density of the muscle, this being about 1060 kg/m³. Accordingly, we have $(0.75 \times 10^5 \text{ J/m}^3)/(1060 \text{ kg/m}^3) = 71$ J/kg. Thus the work that can be expected of a typical muscle in a single slow, strong contraction is about 70 J/kg.

The muscle was taken as having a uniform cross-sectional area. Allowance for tapering may be made by treating the muscle either as a bundle of parallel, cylindrical muscle fibres of different lengths, or as a set of very short lengths of differing cross-sectional area arranged end to end like salami slices. Similar calculations are then performed for each component and the results summed. Either way, the final answer is the same as before.

This value of 70 J/kg is assumed in the calculations that follow, but it is clearly just an estimated typical value. Moreover, tension and work fall the faster a muscle contracts. For example, the faster rate of contraction that is required to maximize the power output may reduce the tension to about 30%, thus lowering the work output to $30/100 \times 70 = 21$ J/kg.

Consider now the 'chin-up', the activity of hanging from a bar by one's hands, then drawing oneself up to touch the bar with the chin. A reason for choosing this activity is that some individuals can do it easily and some hardly at all; for some people one can therefore suppose that the relevant muscles are exerting their maximum tensions. The work required to raise the body is calculated as force multiplied by distance. Raising 1 kg by 1 m requires 9.8 J (i.e. 1 m kg-force = 9.8 J).

9.4.1 During a chin-up, a particular individual's body is lifted 0.4 m. How much work is required *per kg of body mass*?

9.4.2 Assuming that this individual can only just accomplish this action – slowly – and that the relevant muscles perform work at 70 J/kg of muscle, what must the total mass of these muscles be as a percentage of body mass?

For comparison, the total percentage of muscle in a 70-kg man (containing 14% fat) is said to be about 43%. Whether the answer to question 9.4.2 is right for an individual just able to perform a chin-up is hard to check. The relevant muscles dissected from one cadaver (see Notes and Answers) were found to constitute only 1.4% of the body mass, but the man was 80 years old. Considerable amounts of muscle are lost in old age.

Actual body mass is not itself specified in these calculations, but the answers do depend on arm length. Consider now a smaller person with the same bodily proportions.

9.4.3 This smaller person has arms half as long. What percentage of muscle would be required to achieve a chin-up?

There is no implication here concerning the actual muscularity of this person; the conclusion is rather that small people should generally find it easier to lift themselves than do larger ones of comparable build and muscularity (a comforting thought for parents as they compete with their children, perhaps). Another route to this conclusion is given in Notes and Answers.

In high jumping, the advantage is with the taller individual because the centre of gravity starts high, but if jumping is assessed in terms of the raising of the centre of gravity, then tall people have no advantage.

Terrestrial mammals of different sizes tend to jump to similar heights, but the information on athletes, horses, etc. rarely relates to centres of gravity. Two metres is regarded as a suitable height of fence to keep out antelopes, regardless of their size. High jumping as an athletic event involves a run and then a motion of the body that allows the centre of gravity to pass under the bar. Such complexities can be avoided here by considering a standing jump. Rather than looking at a human jump, let us look instead at the more remarkable jump of the bushbaby (*Galago senegalensis*). The results of the calculations are puzzling, but it may be useful at this end of the book to emphasize that calculations can raise questions as well as answer them. Our starting point is a jump recorded for a bushbaby of 0.25 kg – in which the centre of gravity was raised about 7 ft, i.e. 2.1 m.

9.4.4 If a bushbaby leaps upwards, raising its centre of gravity by 2.1 m, how much mechanical work is required per kg of body mass?
(1 m kg-force = 9.8 J.)

9.4.5 On the basis of this answer, and the supposition that the muscles achieve all of the above 70 J/kg (as if a jump involved slow contraction!), how much muscle, expressed as a percentage of body mass, is involved in the jump?

Measurements made on two bushbabies showed that the muscles of the hind legs and back, not all used in jumping, made up 24–25% of the body mass. The total muscle content of the body was 36–37%. What does this mean? How appropriate were the various assumptions?

9.5 Creatine phosphate in muscular contraction

Energy for muscular contraction is provided by the hydrolysis of adenosine triphosphate (ATP). An additional reservoir of energy is provided by creatine phosphate and it is this, rather than the ATP, that is first depleted during muscular activity. The link between creatine phosphate and ATP is through the Lohmann reaction:

$$\text{creatine phosphate} + \text{ADP} = \text{creatine} + \text{ATP}. \tag{9.1}$$

The concentration of creatine phosphate in resting skeletal muscle is commonly 15–25 mmol/kg. This is considerably higher than the concentration of ATP itself, which is nearer 5 mmol/kg. Here we relate the creatine phosphate content of skeletal muscle to work of contraction.

In Section 9.4 a rough value was derived for the amount of work that might be performed by a typical skeletal muscle *in vivo* in a single slow contraction, maximal in terms of tension and shortening. This is 70 J/kg. Since this is expressed per kilogram, like the concentrations of muscle creatine phosphate and ATP, and since the 'energy contents' of these per mole are known, it is tempting to put the information together to see what emerges about the usage of creatine phosphate or ATP per contraction. For example, might all the creatine phosphate be used up in one contraction, hardly any, or some intermediate 'substantial proportion'?

There is no need to consider the (fairly small) free energy change of the Lohmann reaction. Since one molecule of creatine phosphate yields one molecule of ATP, it suffices to consider the free energy change in the hydrolysis of ATP to ADP, ΔG_{ATP}. For cellular conditions, this is given in Section 3.9 as -10 to -13 kcal/mol, or -42 to -54 kJ/mol. Taking it for convenience as -50 kJ/mol, or -50 J/mmol, we have a figure to compare directly with the mechanical work of contraction. In doing this, we may start by assuming complete conversion of chemical energy to external work, and only then give thought to the considerable loss of energy as heat.

....

9.5.1 Assuming completely efficient energy conversion, if a single slow contraction, maximal in terms of tension and of *in vivo* shortening, accomplishes external work to the extent of 70 J/kg and the energy available from ATP is 50 J/mmol, by how much might the concentration of creatine phosphate fall?

....

This answer may be compared with the resting concentration of creatine phosphate of roughly 15–25 mmol/kg muscle. Too much should not be made of the exact answer, however, for the less efficient the actual energy conversion, the more creatine phosphate must be needed per contraction. There is no fixed figure for the percentage efficiency (Section 3.9), but it rarely exceeds 40%. Accordingly the concentration of creatine phosphate should fall by at least 2.5 times the answer to question 9.5.1, namely 3.5 mmol/kg muscle. Remember that the calculations are for a slow contraction; in a strong, rapid contraction against a smaller force the mechanical work is less than 70 J/kg muscle (Section 9.4), so that less creatine phosphate would be used.

Abandoning precise quantification, one can at least say that single strong, slow contractions may use up creatine phosphate to the extent of 'at least several millimoles per kilogram of muscle'.

9.6 Calcium ions and protein filaments in skeletal muscle

Calcium ions play a crucial role in the activation of muscle. In the sarcoplasm of inactive skeletal muscle the concentration of free calcium ions is roughly 10^{-7} mol/l water. Maximum activation requires a rise in concentration to something like 2×10^{-5} mol/l water, at which concentration there is near-maximal binding of calcium to the molecules of troponin-C of the thin filaments. Since there is information readily available on the sizes of filaments and sarcomeres, on the spacing of filaments, and on the numbers of calcium-binding sites on the thin filaments, one may relate these to actual numbers of calcium ions. Some may find the results surprising. (In similar vein, the numbers of hydrogen ions in a mitochondrion were calculated in Section 8.1.)

A skeletal muscle fibre contains a large number of parallel, cylindrical myofibrils, each with a diameter of about 1–2 μm. Within the individual sarcomeres that constitute the myofibrils, the thick filaments are about 1.55 μm long and the thin filaments, anchored to the Z discs, are about 1 μm long. Since it is with the thin filaments that the calcium interacts, let us estimate, for the non-activated state, the number of free calcium ions in that portion of a half-sarcomere that contains thin filaments and is delimited by their ends.

9.6.1 Suppose that the thin filaments at one end of a particular sarcomere are contained within a cylinder of length 1 μm and radius 0.5 μm. What are (a) the cross-sectional area of this cylinder in μm² and (b) the volume in μm³?

Although the cylinder is partially occupied by thick and thin filaments, it may be taken for present purposes as consisting only of water. (The error may be gauged roughly by looking at electron micrographs of muscle. It increases with the degree of overlap between the thick and thin filaments.) To calculate the number of free calcium ions in this volume, recall that 1 mol of any substance contains 6.0×10^{23} molecules (or ions), this being Avogadro's number.

9.6.2 How many free calcium ions are there in that 0.8 μm³ of fluid, if the concentration is 10^{-7} mol/l? (1 μm³ = 10^{-15} l.)

This answer may now be related to the number of thin filaments within this portion of the sarcomere. Seen in transverse section, the parallel thin filaments appear in hexagonal arrays (Figure 9.2). Each thin filament lies in the middle of an equilateral triangle with thick filaments at the angles. The thick filaments are about 45 nm apart. From this information may be calculated first the number of thin filaments per unit area of cross-section and then the number in the cylinder of sarcomere that was specified above.

The area of an equilateral triangle of side S is equal to $\sqrt{3}/4 \times S^2 = 0.43\, S^2$. The area containing a single thin filament is thus $0.43 \times (45\ \text{nm})^2 = 871\ \text{nm}^2 = 8.7 \times 10^{-4}\ \mu\text{m}^2$.

9.6.3 On the basis of this figure, how many thin filaments are there in our cylindrical portion of sarcomere with its cross-sectional area of 0.8 μm²?

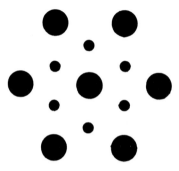

Fig. 9.2. The arrangement of thick and thin filaments in skeletal muscle as seen in transverse section.

In resting muscle this number of thin filaments shares the 0.8 μm³ of sarcomere-end with about 48 free calcium ions (calculated for question 9.6.2).

9.6.4 What is the ratio of thin filaments to free calcium ions?

The point of the series of calculations, apart from exercising quantities that may be already familiar, is just the possible unexpectedness of the final answer. The intended conclusion might seem to be that the high ratio constitutes, in itself, an explanation for the lack of tension in the sarcomere, but that is not so; the number of calcium ions bound by the troponin-C depends also on the affinity of one for the other and this is very high. (If that point is not clear, note that the mechanisms that maintain such low concentrations of calcium ions in the cytosol have also to be responsive to those same concentrations.)

During activation of the muscle the concentration of free calcium ions rises perhaps 200-fold. This means that the number of free calcium ions increases from 48 to $48 \times 200 = 9600$ – call it 10,000. This lowers the ratio of thin filaments to free calcium ions to about 0.1. Even then, there is a large discrepancy between the number of free calcium ions and the number of molecules of troponin-C. These occur in pairs along the thin filament with a spacing of about 37 nm, implying that there are about $1000/37 = 27$ pairs on a 1-μm filament, or about 54 molecules. (This calculation disregards the disruption of the repeating pattern at the filament ends.) Since there are 880 thin filaments in our chosen sarcomere-end, there are about $880 \times 54 = 47,520$ molecules of troponin-C.

On maximal activation, four calcium ions attach to each molecule of tro- ponin-C. We have already estimated that the number of free calcium ions in the cylinder increases to about 10,000.

9.6.5 What then is the ratio of bound calcium to free calcium?

Evidently only a small part of the calcium entering this portion of the sarco- mere on activation remains in free form, a much greater part becoming attached to troponin-C.

The thick filaments were taken as lying about 45 nm apart. Does this square with the appearance of myofibrils in electron micrographs? Seen in longitu- dinal section, the thick filaments span the sarcomeres in numbers that are definitely easily countable, i.e. in tens rather than in hundreds or thousands.

9.6.6 In a myofibril of diameter 1.35 μm (chosen to be 3 × 450 nm), how many thick filaments should lie side by side across the width if their centres are 45 nm apart?

Appendix A

Some useful quantities

Table A.1 *Physical and mathematical quantities*

Temperature (T in K)	$0\,°C = 273.16\,K$
Volume of 1 mol of perfect gas at standard temperature ($0\,°C$) and standard pressure (760 mmHg)	22.4 l
Number of equivalents (equiv) of ionic species of valency z	$z \times$ number of moles
Gas constant (R)	8.31 J/degree mol 1.99 cal/degree mol
Faraday (F) (ionic charge)	23.07 kcal/volt equiv 96,490 coulomb/equiv
e	2.7183
$\ln 10 = 1/\log_{10} e$	2.3026
2.3026 RT/F	at 20 °C $0.0582\,V = 58.2\,mV$ at 37 °C $0.0615\,V = 61.5\,mV$

Table A.2 *Some
atomic and molecular
masses . Units are
'g/mol' (= 'mg/mmol').*

H^+	1.01
Na^+	22.99
K^+	39.10
Mg^{2+}	24.32
Ca^{2+}	40.08
Cl^-	35.46
C	12.01
N	14.01
P	30.98
O	16.00
S	32.07
HCO_3^-	61.02
HPO_4^{2-}	95.99
$H_2PO_4^-$	97.00
SO_4^-	96.07
Glucose	180.18
Lactate$^-$	89.07
Urea	60.06

Appendix B

Exponents and logarithms

Exponents (indices)

The expression x^a means that a quantity, x, is raised to a power a, a being called an exponent or index. The latter may be fractional (e.g. $9^{0.5} = 3$) or negative (e.g. $4^{-1} = 0.25$). Whatever the value of x, $x^0 = 1$. Expressions involving more than one exponent may be manipulated according to the following rules:

1. $x^a \times x^b \quad = x^{(a + b)}$.
2. $x^a / x^b \quad = x^{(a - b)}$.
3. $(x^a)^b \quad = x^{a\,b}$.

Rules 2 and 3 follow from the first. As a special case of rule 2, $1/x^b = x^{-b}$. Although a and b often stand for numbers, they may also stand for variables or more complicated expressions, as in equation 1.7.

Introduction to logarithms

Looking at graphs with logarithmic scales, most people make some sort of sense of them even when logarithms themselves are unfamiliar. Let us therefore take such scales as a starting point. As explained in Section 1.5, they are commonly used to plot quantities that vary so hugely as to fit uncomfortably on linear scales. A logarithmic scale may look like this:

$10^0 \qquad\qquad 10^1 \qquad\qquad 10^2 \qquad\qquad 10^3 \qquad\qquad 10^4 \rightarrow$ etc.

On this scale, the equally spaced ticks correspond to tenfold increases in the numbers plotted. The numbers are expressed as 10^0, 10^1, etc. (rather than as 1, 10, 100, etc.) partly for convenience, for extension of the scale to the right soon leads to numbers that otherwise take up too much space (e.g. 1,000,000,000 for 10^9). The scale may also be extended to the left to accommodate numbers less than 1, i.e. with 10^{-1} and 10^{-2} for 0.1 and 0.01 respectively. There is never a zero on a logarithmic scale.

The numerals indicating those powers of ten are exponents, but they can also be called the 'logarithms' of the numbers. Thus the logarithm of 10,000, i.e. 10^4, is

4. More precisely, 4 may be described as the 'common logarithm' of 10^4, or its logarithm 'to the base 10', for other numbers may be chosen as base (see below). In symbols, $\log_{10}(10^4) = 4$, with the subscript denoting the base. Conventionally, '$\log_{10}$' is abbreviated to 'log' when there is no risk of confusion. Thus, for the general case of a number a, we have:

$$\log(10^a) = a.$$

Because $1 = 10^0$ and $10 = 10^1$, then $\log 1 = 0$ and $\log 10 = 1$. Negative values of a are also possible, as already noted.

Another way of expressing the relationship between a and $\log a$ is this:

$$10^{\log a} = a.$$

This is the key to converting $\log a$ to a using a calculator. Many calculators have a button marked 10^x which converts x (or a) to its 'antilogarithm', i.e. the number of which the logarithm is x.

We can now establish how to plot on the above scale any values that are intermediate between 10^0 and 10^1, between 10^1 and 10^2, etc. Let us choose 3 and 30. The calculator tells us that $\log 3$ is 0.477 and $\log 30$ is 1.477 (roughly 0.5 and 1.5). We can therefore locate 3 and 30 almost midway between 10^0 and 10^1 and between 10^1 and 10^2, respectively:

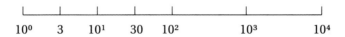

10^0 3 10^1 30 10^2 10^3 10^4

Graph paper is available that has logarithmic scales for one or both axes. However, it is often more convenient, and usually more accurate, to use ordinary (linear) graph paper and convert numbers to logarithms before plotting. Then, there is a zero to the scale (even if it is not actually plotted) and numbers less than one become negative.

Rules for manipulating logarithms

The rules relating to exponents lead to analogous rules for logarithms:

1. **$\log (ab) = \log a + \log b$.**
2. **$\log (a/b) = \log a - \log b$.**

As a special case of the latter equation:

$$\log (1/b) = \log 1 - \log b = 0 - \log b = -\log b.$$

Thus $\log 0.1 = -1$ and $\log 0.001 = -3$.

3. **$\log (b^a) = a \log b$.**

The relationship given earlier, that $\log(10^a) = a$, is consistent with rule 3, since $\log 10 = 1$. At one time these rules, together with tables of logarithms, were as important to practical multiplication and division as calculators are now. They are applied to numbers in the following paragraphs, but they are also useful in the algebraic manipulation of formulae.

A logarithm to remember

Few of us need to remember the numerical values of logarithms (beyond those for 0.1, 1, 10, 100, etc.). However, it can really be worthwhile to remember at least one. With this one logarithm, one may calculate others, perform many physiological calculations without recourse to a calculator, and appear more of a mathematician than perhaps one is.

The recommended logarithm is $\log_{10} 2$. To five significant figures, it is memorable for its symmetry, being **0.301030**, but, for physiological calculations, 0.30 is usually close enough. This is applied a number of times in the main part of the book.

When a logarithm is needed, the odds of it being that particular one might seem to be slim, but in a moment we will see how a few other logarithms may be quickly derived from it. Moreover, when one performs approximate calculations in physiology, one is often free to choose the quantities involved in order to simplify the arithmetic.

The remainder of this Appendix is about how to derive certain logarithms easily when one is parted from logarithm tables and calculator. Some people find aesthetic pleasure in such things; those who just find them stressful may yet care to proceed further, as an exercise. The calculations are detailed in Notes and Answers.

....

1. Computer users may recall that a kilobyte is not exactly 1000 bytes, but 1024 bytes. 1024 is 2^{10} and it is nearly equal to 10^3. Taking 2^{10} as identical to 10^3, what is log 2?

....

Armed with the logarithm of 2, let us now see how the logarithms of some other whole numbers may be obtained from it. To find log 4, note that $4 = 2^2$. It follows that log 4 is twice log 2 and is therefore calculable as $2 \times 0.301 = 0.602$.

....

2. Likewise, since $8 = 2^3$, $\log 8 = 3 \times \log 2$. What is log 8?

3. Since $5 = 10/2$, $\log 5 = (\log 10 - \log 2)$. What is log 5?

4. The square of 7 is very nearly half of 100. Log 7 is thus only slightly (0.5%) less than half of ($\log 100 - \log 2$). What is that?

....

Thus far, we have the logarithms of 1, 2, 4, 5, 7, 8 and 10, and Table B.1 gives these values in bold type. This may be far enough to take the process in the context of physiological mental arithmetic, but it is easy enough to add others in the same

Table B.1. *Some logarithms*
(to base 10)

x	$\log x$
1	**0**
1.6	*0.204*
2	**0.301**
2.5	*0.398*
3	0.477
3.5	*0.544*
4	**0.602**
5	**0.699**
6	0.778
7	**0.845**
8	**0.903**
9	0.954
10	**1.000**

Values shown **bold** are
those most simply
obtained in the way
described in the text.
Those that are *italicized*
are obtainable from the
bold values, making
further use of log 2.

way. Thus, x may be taken as 5/2, 7/2, (8 × 2/10) and so on. These three values are italicized in the table.

Those wishing to estimate log 3 by comparable means could start from the fact that 3^4 (= 81) is close to 10 × 2^3 (= 80). Approximately, therefore, 4 log 3 = (log 10 + 3 log 2) = 1.903. Hence, log 3 = 1.903/4 = 0.476 (the true value being 0.477). To proceed further, doubling log 3 gives log 9, and adding log 2 to log 3 gives log 6.

Given all these logarithms, one may prepare a much fuller logarithmic scale than either of those depicted above. Even the few logarithms most easily derived from log 2 suffice for rough exploratory plots, and to illustrate again the nonlinearity of logarithmic scales:

1   2   4   5   8   10

This time there is just a single 'cycle', from 1 to 10, but the spacing for other cycles is the same (e.g. for 20, 40, 50, 80, 100). The unnumbered vertical ticks, equally spaced like the lines on ordinary linear graph paper, are ignored once the scale is prepared.

Precision and accuracy

What is the implication of taking log 2 as 0.30 instead of 0.301? The antilogarithm of 0.300 ($10^{0.300}$) is 1.9953 instead of 2.0000 – a difference of 0.23%. Generally, a difference in log x of 0.001 implies a difference in x of 0.23%. A difference in log x of 0.01 implies a difference in x of 2.3%.

When logarithmic scales are used for graphing, large numbers are compressed together and the differences between small ones are emphasized. This can have an important effect on how we perceive them – notably in relation to errors in the smaller values. For example, suppose that one measures lengths of 0–100 mm with an ordinary ruler that is marked in millimetres, with measurements less than 1 mm being proportionately the least accurate. On a logarithmic scale, measurements of 0.05 and 0.1 mm, of dubious accuracy, have as much prominence as accurate values of 50 and 100 mm. If such data were plotted on a graph, and a line were fitted to the resulting points, then the lower values should not be allowed to have an undue effect on the placing of that line.

Natural logarithms

The logarithms described above are of the kind most commonly and conveniently used by most of us, i.e. common logarithms 'to the base 10'. Other numbers may be used as the base instead of 10, most notably the irrational number called e, which has a value approximating to 2.718. Logarithms to this base are known as natural, or Napierian, logarithms, and are preferred by mathematicians because they relate most naturally to other aspects of mathematics. The natural logarithm of x is written as '$\log_e x$' or as 'ln x'. The rules for manipulating natural logarithms are the same as already described for common logarithms. One form of logarithm may be converted to the other by means of the following relationships (in which the '0.4343' is both $\log_{10} e$ and $1/\log_e 10$, and the 2.3026 is its reciprocal):

$$\log_{10} x = 0.4343 \times \log_e x.$$
$$\log_e x = 2.3026 \times \log_{10} x.$$

REFERENCES

Agostoni E. & D'Angelo E. (1991) Pleural liquid pressure. *Journal of Applied Physiology*, **71**, 393–403.

Alexander R.McN. (1992) The work that muscles can do. *Nature*, **357**, 360–361.

Bangham A.D. (1991) Pattle's bubbles and Von Neergaard's lung. *Medical Science Research*, **19**, 795–799.

Bence-Jones H. (1845) Contributions to the chemistry of urine. On the variation in the alkaline and earthy phosphates in the healthy state, and on the alkalescence of urine from fixed alkali. *Philosophical Transactions of the Royal Society B*, **135**, 335–349.

Blaxter K. (1989) *Energy Metabolism in Animals and Man.* Cambridge: Cambridge University Press.

Bowman W.C. & Rand M.J. (1980) *Textbook of Pharmacology*, 2nd edn. Oxford: Blackwell Scientific Publications.

Brockway J.M. (1987) Derivation of formulae used to calculate energy expenditure in man. *Human Nutrition: Clinical Nutrition*, **41C**, 463–471.

Burton A.C. (1965) *Physiology and Biophysics of the Circulation.* Chicago: Year Book Medical Publishers Inc.

Burton R.F. (1973) The significance of ionic concentrations in the internal media of animals. *Biological Reviews*, **48**, 195–231.

Burton R.F. (1987) On calculating concentrations of 'HCO$_3$' from pH and PCO$_2$. *Comparative Biochemistry and Physiology*, **87A**, 417–422.

Burton R.F. (1988) The protein content of extracellular fluids and its relevance to the study of ionic regulation: net charge and colloid osmotic pressure. *Comparative Biochemistry and Physiology*, **90A**, 11–16.

Burton R.F. (1992) The roles of intracellular buffers and bone mineral in the regulation of acid–base balance in mammals. *Comparative Biochemistry and Physiology*, **102A**, 425–432.

Burton R.F. (1998) *Biology by Numbers.* Cambridge: Cambridge University Press.

Chacon E., Reece J.M., Nieminen A.-L., Zahrebelski G., Herman B. & Lemasters J.J. (1994) Distribution of electrical potential, pH, free Ca^{2+}, and volume inside cultured adult rabbit cardiac myocytes during chemical hypoxia – a multiparameter digitized confocal microscopic study. *Biophysical Journal*, **66**, 942–952.

Clausen T., Van Hardeveld C. & Everts M.E. (1991) Significance of cation transport in control of energy metabolism and thermogenesis. *Physiological Reviews,* **71,** 733–774.

Cohn S.H., Vaswani A., Zanzi I & Aloia J.F. (1976) Changes in body chemical composition with age measured by total-body neutron activation. *Metabolism,* **25,** 85–103.

Diem K. (1962) *Documenta Geigy. Scientific Tables,* 6th edn. Manchester: Geigy Pharmaceutical Co. Ltd.

Driessens F.C.M., van Dijk J.W.E. & Verbeeck R.M.H. (1986) The role of bone mineral in calcium and phosphate homeostasis. *Bulletin des Sociétés chimiques Belges,* **95,** 337–342.

Driessens F.C.M., Verbeeck R.M.H. & van Dijk J.W.E. (1989) Plasma calcium difference between man and vertebrates. *Comparative Biochemistry and Physiology,* **93A,** 651–654.

Durnin J.V.G.A. & Passmore R. (1967) *Energy, Work and Leisure.* Heinemann Educational Books Ltd.

Edwards N.A. (1975) Scaling of renal functions in mammals. *Comparative Biochemistry and Physiology,* **52A,** 63–66.

Galilei G. (1637) *Dialogues Concerning Two New Sciences* (translated by H. Crew & A. De Salvio 1914). London: Macmillan.

Gil J., Bachofen H., Gehr P. & Weibel E.R. (1979) Alveolar volume-surface area relation in air- and saline-filled lungs fixed by vascular perfusion. *Journal of Applied Physiology,* **47,** 990–1001.

Green J. & Kleeman C.R. (1991) Role of bone in regulation of systemic acid–base balance. *Kidney International,* **39,** 9–26.

Guyton A.C., Moffatt D.S. & Thomas H.A. (1984) Role of alveolar surface tension in transepithelial movement of fluid. In *Pulmonary Surfactant.* (eds B. Robertson, L.M.G. van Golde & J.J. Batenburg) pp. 171–185. Amsterdam: Elsevier.

Harvey R.J. (1974) *The Kidneys and the Internal Environment.* London: Chapman and Hall.

Harvey W. (1628) *Exercitatio Anatomica de Motu Cordis et Sanguinis in Animalibus.* Frankfurt: Guilielmi Fitzeri.

Harwood P.D. (1963) Therapeutic dosage in small and large mammals. *Science,* **139,** 684–685.

Hedrick M.S. & Duffield D.A. (1986) Blood viscosity and optimal hematocrit in a deep-diving mammal, the northern elephant seal (*Mirounga angustirostris*). *Canadian Journal of Zoology,* **64,** 2081–2085.

Hilden T. & Svendsen T.L. (1975) Electrolyte disturbances in beer drinkers. *Lancet,* **2,** 245–246.

Hofstadter D. (1985) On number numbness. In *Metamagical Themas,* pp 115–135. Harmondsworth: Viking, Penguin Books Ltd.

Johnson C.D., Mole D.R. & Pestridge A. (1995) Postprandial alkaline tide: does it exist? *Digestion,* **56,** 100–106.

Keys A. & Brožek J. (1953) Body fat in adult man. *Physiological Reviews*, 33, 245–325.

Kleiber M. (1961) *The Fire of Life*. New York: John Wiley and Sons.

Lavoisier A. L. & Laplace P. S. (1780) Mémoire sur la chaleur. In *OEuvres de Lavoisier: publiées par les soins de Son Excellence le Ministre de l'Instruction Publiques et des Cultes*. Volume 2 (1862). *Mémoires de Chimie et de Physique*, pp. 283–333. Paris: Imprimerie Impériale.

Lentner C. (1981) *Geigy Scientific Tables. Volume 1. Units of Measurement, Body Fluids, Composition of the Body, Nutrition*. Basle: Ciba-Geigy Ltd.

Lindstedt S.L. & Calder W.A. (1981) Body size, physiological time, and longevity of homeothermic animals. *Quarterly Review of Biology*, 56, 1–16.

Ludwig K. (1844) Nieren und Harnbereitung. In *Handwörterbuch der Physiologie* (ed Wagner.) pp. 628–640. Brunswick: Vieweg.

Mandelbrot B.B. (1983) *The Fractal Geometry of Nature*. New York: Freeman.

Peters R.H. (1983) *The Ecological Implications of Body Size*. Cambridge: Cambridge University Press.

Pfaller W. & Rittinger M. (1980) Quantitative morphology of the rat kidney. *International Journal of Biochemistry*, 12, 17–22.

Pirofsky B. (1953) The determination of blood viscosity in Man by a method based on Poiseuille's law. *Journal of Clinical Investigation*, 32, 292–298.

Prange H.D., Anderson J.F. & Rahn H. (1979) Scaling of skeletal mass to body mass in birds and mammals. *American Naturalist*, 113, 103–122.

Rahn H. (1966) Aquatic gas exchange: Theory. *Respiration Physiology*, 1, 1–12.

Robertson G.L., Shelton R.L. & Athar S. (1976) The osmoregulation of vasopressin. *Kidney International*, 10, 25–37.

Roy D.R., Layton H.E. & Jamison R.L. (1992) Countercurrent mechanism and its regulation. In *The Kidney: Physiology and Pathophysiology*, 2nd edn (eds D. W. Seldin & G. Giebisch) pp. 1649–1692. New York: Raven Press.

Rubner M. (1883) Über den Einfluss der Körpergrösse auf Stoff- und Kraftwechsel. *Zeitschrift für Biologie*, 19, 535–562.

Sarrus & Rameaux (1838–9) *Bulletin de l'Académie Royale de Médicine*, 3, 1094–1100.

Schafer J.A. (1982) Salt and water absorption in the proximal tubule. *The Physiologist*, 25, 95–103.

Schmidt-Nielsen K. (1984) *Scaling: Why is Animal Size so Important?* Cambridge: Cambridge University Press.

Schwerzmann K., Hoppeler H., Kayar S.R. & Weibel E.R. (1989) Oxidative capacity of muscle and mitochondria: correlation of physiological, biochemical, and morphometric characteristics. *Proceedings of the National Academy of Sciences, USA*, 86, 1583–1587.

Stahl W.R. (1965) Organ weights in primates and other mammals. *Science*, 150, 1039–1042.

Stone H.O., Thompson H.K. & Schmidt-Nielsen K. (1968) Influence of erythrocytes on blood viscosity. *American Journal of Physiology*, 214, 913–918.

Thaysen J.H., Lassen N.A. & Munck O. (1961) Sodium transport and oxygen consumption in the mammalian kidney. *Nature*, **190**, 919–921.

Wells R.M.G. & Baldwin J. (1994) Oxygen transport in marine green turtle (*Chelonia mydas*) hatchlings: blood viscosity and control of hemoglobin oxygen affinity. *Journal of Experimental Biology*, **188**, 103–114.

West L.J., Pierce C.M. & Thomas W.D. (1962) Lysergic acid diethylamide: its effect on a male Asiatic elephant. *Science*, **138**, 1100–1102.

Zeuthen T. & Stein W.D. (1994) Cotransport of salt and water in membrane proteins: membrane proteins as osmotic engines. *Journal of Membrane Biology*, **137**, 179–195.

(numbered according to section)

1 Introduction to physiological calculation – approximation and units

1.3 How attention to units can ease calculations, prevent mistakes and provide a check on formulae

1.3A

If, in the urea calculation, the initial data were given as 65 ml/h and 230 mmol/l, then one could begin either by re-expressing the first as 0.065 l/h or by re-expressing the second as 0.23 mmol/ml (as actually presented). Then, litres would cancel with litres in the calculation, or ml would cancel with ml. Likewise, when calories meet joules, mmHg meets kilopascals, etc., one might start by applying appropriate conversion factors as in Table 1.1. However, in order to calculate with units in an overtly orderly fashion, one may choose to follow a different procedure. For the above examples, we have: $1\,l = 1000$ ml, 1 cal $= 4.1855$ J and 1 mmHg $= 0.1333$ kPa. Each of these equations can be rearranged to yield an expression equal to 1: $1000\,\dfrac{\text{ml}}{\text{l}} = 1$, $4.1855\,\dfrac{\text{J}}{\text{cal}} = 1$ and $0.1333\,\dfrac{\text{kPa}}{\text{mmHg}} = 1$. Since any quantity is unchanged when it is multiplied by 1, it is also, in a sense, unchanged when it is multiplied by one of these expressions. Therefore, any one of them may be introduced into a calculation, as appropriate, in order to rationalize the units used. As an example, suppose that the urea calculation has reached the following seemingly inappropriate point, with two different units for urine volume.

Rate of urea excretion $= 65$ ml/h $\times 230$ mmol/l.

The right-hand expression is now divided by 1000 ml/l (which equals 1):

$$\text{rate of urea excretion} = 65 \text{ ml/h} \times 230 \text{ mmol/l} \div 1000 \text{ ml/l}$$
$$= 15 \text{ mmol/h.}$$

1.3B

Dimensional analysis resembles unit analysis, but it uses less information and has a more limited role. Instead of writing, say, kg, m and s, one writes M, L and T, standing for mass, length and time, respectively. These are combined, for example, to give L^2, L^3, L/T for area, volume and velocity respectively, and to give ML/T^2, M/LT^2 and ML^2/T^2 for force, pressure and energy. The letters S, A, K and I are used, respectively, for molar amount of substance, electric current, temperature and luminous intensity.

1.3C

The diffusion of gases may be treated in terms of partial pressures instead of concentrations. For oxygen, equations 1.4 and 1.5 are combined in the following:

$$\text{rate} = (S_{O_2} P_{1O_2} - S_{O_2} P_{2O_2}) \times a/d \times D$$
$$= (P_{1O_2} - P_{2O_2}) \times a/d \times (DS_{O_2}).$$

The expression (DS_{O_2}) can be regarded as a single entity, a diffusion constant that is distinct from D. Appropriate units could be mmol/(mmHg cm s). This approach is used in Section 5.3.

Practice in unit analysis

1. Equation 1.6 rearranges to:

$$\text{viscosity} \propto \text{pressure difference} \times \frac{\text{radius}^4}{\text{length} \times \text{flow rate}}$$

Therefore, units for viscosity $= \dfrac{N}{m^2} \times \dfrac{m^4}{m \times m^3/s}$

$$= N\,s/m^2.$$

In practice, viscosities are often given in centipoises, where 1 poise $= 0.1$ Ns/m^2. Conveniently, the viscosity of water at 20 °C is about 1 centipoise.

2. $E = mc^2$: E is energy, m is mass (kg), c is the velocity of light (m/s). Therefore, the units for mc^2 are kg m^2/s^2, but since 1 N $=$ 1 kg m/s^2, kg m^2/s$^2 =$ Nm. Finally, Nm $=$ J $=$ unit of energy.

3. According to Appendix A, $R = 1.99$ cal/degree mol and $F = 23.07$ kcal/volt equiv. T is in Kelvin (K). The number of equivalents of a substance is the number of moles multiplied by the valency, z. In units, RT/zF therefore becomes:

$$\frac{\text{cal}}{\text{degree} \times \text{mol}} \times \frac{\text{degree}}{z} \times \frac{\text{volt} \times (z \times \text{mol})}{\text{cal}} = \text{volt}.$$

4. Since the period increases with pendulum length (L) and decreases with g, test first the hypothesis that it is simply proportional to L/g. In terms of SI units, this is m/(m/s^2) $=$ s^2, rather than s, and is therefore wrong. The problem is solved by taking the square root of L/g. At this point the correct formula might be remembered as '$2\pi\sqrt{(L/g)}$', but the '2π' cannot be obtained by unit analysis.

1.4 Analysis of units in expressions involving exponents (indices)

An **exponential time course** in accordance with equation 1.7 applies whenever the variable, Y, changes at a rate that is proportional to its own instantaneous value. (In the language of calculus, $Y = Y_0 e^{kt}$ when $dY/dt = kY$.) Suppose, for example, that a drug present only in the blood plasma is excreted by the kidneys at a rate that is proportional to the plasma concentration (as would be the case if the drug were filtered by the glomeruli and neither secreted nor reabsorbed by the tubules). Then, if no other process were involved to complicate the picture, the concentration of the drug in the plasma would fall exponentially (Section 6.3).

1.5 Logarithms

Equation 1.11 is derived from the relationship between the dissociation constant, K, and the equilibrium concentrations of reactants in the overall reaction (i.e. ignoring the intermediate production of carbonic acid, which exists in far smaller amounts than carbon dioxide and bicarbonate):

$$CO_2 + H_2O = H_2CO_3 = H^+ + HCO_3^-.$$

Because bicarbonate dissociates further to produce carbonate (Section 8.6), K for this first dissociation is denoted K_1. In dealing with concentrations rather than with activities, K_1 is further characterized as K_1'. It relates to the concentrations (ignoring that of water, which is conventionally taken as unity) as follows:

$$K_1' = [H][HCO_3]/[CO_2].$$

Taking logarithms of both sides, we have:

$$\log K_1' = \log([H][HCO_3]/[CO_2])$$
$$= \log[H] + \log([HCO_3]/[CO_2]).$$

If $\log[H^+]$ is taken as $-pH$ (but see Section 8.1) and $\log K_1'$ is defined similarly as $-pK_1'$, then this equation can be rearranged to give equation 1.11.

2 Quantifying the body: interrelationships amongst 'representative' or 'textbook' quantities

2A

Thinking to provide here a conventional definition of the 'standard 70-kg man', I found him surprisingly elusive. All the physiologists I asked recalled meeting him, but could not help me to pin him down. I found him (albeit undefined) in only one of the many textbooks I looked at and, although men of about 70 kg are mentioned in some nineteenth-century textbooks, these men are not 'standard'. Keys and Brožek (1953) refer to, 'our standard "normal young man"' containing 61% water, 14% fat, 19% cell solids, 6% bone mineral, 64% cells and 16% extracellular fluid. Supplement 6 to the *British Journal of Radiology* (1955) describes a 'standard man' of 70 kg intended to provide a common basis for calculations in radiological laboratories.

With regard to mundane quantitative data not represented in this chapter, e.g. typical rates at which particular substances are excreted, it often pays to look in older textbooks, for the newer ones need more space for recent discoveries. Compilations of data, like those of Diem (1962) and Lentner (1981), are invaluable.

2B

The *extracellular fluid volume*, for which a single representative value is given, is not only variable from person to person like all the other quantities, but depends on how it is defined and measured. Apart from the blood plasma most of the extracellular compartment consists of largely unambiguous interstitial fluid amongst the cells in general, but there are specialized fractions (transcellular fluid) such as the cerebrospinal, synovial and intraocular fluids and the fluid in bone matrix. Extracellular fluid volume is usually best defined in terms of the method of measurement. The general method involves infusing a marker substance into the blood and allowing it to equilibrate throughout the supposed extracellular fluid. The amount of marker present at equilibrium is taken as the quantity infused, but corrected for losses due to excretion. The extracellular volume is then calculated as the amount present divided by the concentration at the same moment. (This 'dilution principle' is also used, with appropriate markers, for estimating other volumes, such as those of blood plasma and total body water.) Many marker substances have been used, notably inulin, sucrose, mannitol, thiosulphate, radiosulphate and the complex of chromium and ethylenediaminetetra-acetic acid (EDTA). They give slightly different results, according to where exactly they penetrate (even within a muscle, for example), so what they measure is most accurately described as 'inulin space', 'sucrose space' and so on. In the context of natural variation and our representative 14 l, these discrepancies are of little consequence. The intracellular volume of water is estimated by subtracting the extracellular volume from the total water content of the body and is therefore no more precisely defined.

2.1 $150 \text{ mmHg} - 760 \text{ mmHg} \times (250\text{ml}/\text{min})/(3795 \text{ ml}/\text{min}) = 100 \text{ mmHg}.$
2.2 $40 \text{ mmHg}/50 \text{ mmHg} = 0.8$ as before.

3 Energy and metabolism

3.1 Measures of energy

The values in Table 3.1 differ slightly from those in the first edition. Various sources have been consulted, including Brockway (1987) and Blaxter (1989).

3.1.1 $100 \text{ kg} \times 0.8 \text{ kcal}/(\text{kg °C}) \times 1\text{°C} = 80 \text{ kcal}.$
3.1.2   (a) $100 \text{ kg-force} \times 10 \text{ m} \times 9.8 \text{ J}/\text{m kg-force} \div 1000 \text{ kJ}/\text{J} = 9.8 \text{ kJ},$
         (b) $9.8 \text{ kJ} \div 4.19 \text{ kJ}/\text{kcal} = 2.3 \text{ kcal}.$

3.2 Energy in food and food reserves; relationships between energy and oxygen consumption

3.2.1 $10 \text{ kg} \times 6.6 - 10 \text{ kg} = 56 \text{ kg}$.

3.2.2   4.8 differs from both 5.0 and 4.6 by 0.2; $0.2/4.8 \times 100 = 4.2\%$.

For the calculation of energy expenditure from measurements of oxygen, carbon dioxide and urinary nitrogen, see Brockway (1987).

3.3 Basal metabolic rate

3.3.1   (a) $1700 \text{ kcal/day} \div 4.2 \text{ kcal/g} = 405 \text{ g/day}$, or $7000 \text{ kJ/day} \div 17.5 \text{ kJ/g} = 400 \text{ g/day}$.
         (b) $1700 \text{ kcal/day} \div 9.3 \text{ kcal/g} = 183 \text{ g/day}$, or $7000 \text{ kJ/day} \div 39.1 \text{ kJ/g} = 179 \text{ g/day}$.

3.3.2 $8640 \text{ kJ/day} \times 1000 \text{ J/kJ} \div 86{,}400 \text{ s/day} = 100 \text{ J/s} = 100 \text{ W}$.

3.4 Oxygen in a small dark cell

3.4.1 $1600 \text{ l} \div 16 \text{ l/h} = 100 \text{ h}$.

3.4.2   Half of the original percentage of oxygen in the air = ca. 10%.

3.4.3   10% of 760 mmHg = 76 mmHg.

3.5 Energy costs of walking, and of being a student

3.5.1   (a) $240 - 90 = 150 \text{ kcal}$, or $1000 - 377 = 623 \text{ kJ}$.
         (b) $150 \text{ kcal} \div 3.7 \text{ kcal/g} = 40.5 \text{ g}$, or $623 \text{ kJ} \div 15.6 \text{ kJ/g} = 40 \text{ g}$.

3.5.2 $450/150 = 3 \text{ h}$.

Energy requirements for many activities and walks of life are given by Durnin & Passmore (1967).

3.6 Fat storage and the control of appetite

3.6.1 $93 \text{ kcal} \div 9.3 \text{ kcal/g} = 10 \text{ g}$, or $389 \text{ kJ} \div 39.1 \text{ kJ/g} = 10 \text{ g}$.

3.6.2 $10 \text{ g/day} \times 365 \text{ day} = 3650 \text{ g}$.

3.7 Cold drinks, hot drinks, temperature regulation

3.7.1 $25\,°C \times 0.60 \text{ kg}/60 \text{ kg} = 0.25\,°C$.

3.7.2 $25\,°C \times 0.6 \text{ kg} \times (1 \text{ kcal/kg °C}) = 15 \text{ kcal}$.

3.7.3 $(15 \text{ kcal})/(100 \text{ kcal/h}) = 0.15 \text{ h}$.

3.7.4 $(1.67 \text{ kcal/kg h})/(0.8 \text{ kcal/kg °C}) = 2.1\,°C$.

3.7.5 $(100 \text{ kcal/h})/(0.58 \text{ kcal/g}) = 172 \text{ g/h}$.

3.8 Oxygen and glucose in blood

3.8.1 $200 \text{ ml/l} \div 22.4 \text{ ml/mmol} = 9 \text{ mmol/l}$.

3.8.2 $900 \text{ mg/l} \div 180 \text{ mg/mmol} = 5 \text{ mmol/l}$.

3.8.3 $2 \text{ mmol/(l h)} \times 45/100 = 0.9 \text{ mmol/(l h)}$.

3.8.4   (a) $0.9 \text{ mmol/(l h)} \times 5 \text{ l} \times 24 \text{ h/day} = 108 \text{ mmol/day}$,
         (b) $108 \text{ mmol/day} \div 1000 \text{ mmol/mol} \times 180 \text{ g/mol} = 19 \text{ g/day}$.

3.9 Adenosine triphosphate and metabolic efficiency

3.9.1   38 mol $\times$ 11 kcal/mol = 418 kcal, or 38 mol $\times$ 46 kJ/mol = 1748 kJ.

3.9.2   418 kcal/670 kcal $\times$ 100 = 62%, or 1748 kJ/2800 kJ/mol $\times$ 100 = 62%.

3.9.3 $3/(14 - 1.4) \times 100 = 23.8\%$.

3.9.4   9.8 J/30 J $\times$ 100 = 33%.

3.9.5   686 J/3480 J $\times$ 100 = 19.7%.

3.10 Basal metabolic rate in relation to body size

3.10.1   The man: 1700 kcal/day $\div$ 70 kg = 24 kcal/kg day, or 7000 kJ/day $\div$ 70 kg = 100 kJ/kg day.
The mouse: 4.8 kcal/day $\div$ 0.03 kg = 160 kcal/kg day, or 20 kJ/day $\div$ 0.03 kg = 667 kJ/kg day.

3.10.2   If the small mammal has mass M, surface area S and specific metabolic rate m, then heat production = Mm = heat loss = $k(37\,°C - 17\,°C) \times S$, where k is a constant independent of size. Correspondingly, for the large mammal, with its unknown body temperature denoted x, 1000 Mm = $k(x\,°C - 17\,°C) \times 100\ S$. Thus Mm/kS is equal both to $(37\,°C - 17\,°C)$ and to $100(x\,°C - 17\,°C)/1000$. Equating these, and rearranging, we have $x = 17 + (37 - 17) \times 10 = 217\,°C$. Compare this calculation with that of Kleiber (1961) that is mentioned in 'Preface to the First Edition'.

Equation 3.3: for more on this, and on the propriety of taking logarithms here, see Sections 1.4 and 1.5 and an associated Note.

3.10.3   70 kcal/day $\times$ 24.2 = 1694 kcal/day,
or 293 kJ/day $\times$ 24.2 = 7091 kJ/day.

These are within the ranges given in Section 3.3.

3.10.4   70 kcal/day $\times$ 5.6 = 392 kcal/day,
or 293 kJ/day $\times$ 5.6 = 1641 kJ/day.

3.10.5   70 kcal/day $\times$ 5623 $\div$ 100,000 kg = 3.9 kcal/day,
or 293 kJ/day $\times$ 5623 $\div$ 100,000 kg = 16.5 kJ/kg day.

Given the logarithmic relationship of equation 3.4, and the huge range of mammalian body sizes, it is useful to plot specific BMR and body mass using logarithmic axes (see Chapter 1). The data for human, mouse and whale (questions 3.10.1 and 3.10.5) are shown in Figure 1.
For more on scaling and body size, see Schmidt-Nielsen (1984), Peters (1983) and Burton (1998).

3.11 Drug dosage and body size

3.11.1   1700A/4.8 = 354A, or 7000A/20 = 350A.

3.11.2   2333A/354 (or 2333A/350A) = nearly 7.

For the sad story of the elephant, and comments on the dosage, see West, Pierce & Thomas (1962) and Harwood (1963).

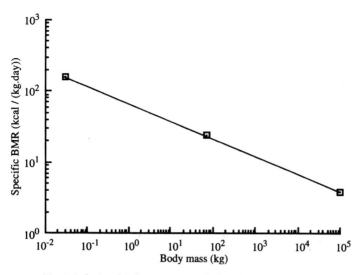

Fig. 1. Relationship between specific basal metabolic rate (specific BMR) and body mass: data for a man and a mouse (question 3.10.1) and for a whale (question 3.10.5). Note the logarithmic scales.

3.12 Further aspects of allometry – life span and the heart

The problem that could arise here in raising M to a power, given that M has units (kg), is discussed in Section 1.4.

3.12.1   Number of beats $= 241\ M^{-0.25} \times 6 \times 10^6\ M^{+0.20} = 1.4 \times 10^9\ M^{-0.05}$.

3.12.2 $1^{-0.05} = 1.0$.

3.12.3 $241 \times 0.35 = 84$ beats/min.

3.12.4 $11.8 \times 2.3 = 27$ years.

Body size, physiological time and longevity are discussed by Lindstedt & Calder (1981).

3.13 The contribution of sodium transport to metabolic rate

The 'recent estimate' that 20% of the resting metabolic rate is used for sodium transport is given by Clausen, Van Hardeveld & Everts (1991).

3.13.1   20% of 24 = 4.8 kcal/kg day, or 20% of 100 = 20 kJ/kg day.

3.13.2   4.8 kcal/kg day ÷ 3.9 kcal/kg day × 100 = 123%, or 20 kJ/kg day ÷ 16.5 kJ/kg day × 100 = 121%.

3.14 Production of metabolic water in human and mouse

3.14.1   16 mol/day × 18 g/mol = 288 g/day = 288 ml/day.

3.14.2   400 ml/day × 7 = 2800 ml/day.

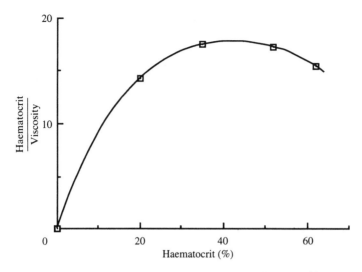

Fig. 2. Data of question 4.2.1, showing that the ratio of haematocrit (percentage) to viscosity (centipoise) is maximal (and optimal) at a near-normal value of the haematocrit.

4 The cardiovascular system

4.1 Erythrocytes and haematocrit (packed cell volume)

4.1.1 $0.91 \times 44 = 40\%$ exactly!

4.1.2   Box volume $= 60 \ \mu m^2 \times 2.4 \ \mu m = 144 \ \mu m^3$; $84 \ \mu m^3$ is 58% of this. The calculation is adapted from A.C. Burton (1965).

4.1.3 $1 \ mm^3 \times 44/100 \div (5.4 \times 10^6) \times 10^9 \ \mu m^3/mm^3 = 81.5 \ \mu m^3$.

4.1.4 $157 \ \mu m^3/84 \ \mu m^3 = 1.87$.

4.1.5 $(156 - 30) \times \%$ NaCl $= 48.6$. So % NaCl $= (84 - 30) \times 0.9/(156 - 30) = 0.39\%$.

4.1.6 $84 \ \mu m^3 + 180 \ \mu m^3 = 264 \ \mu m^3$.

4.1.7 $84 \ \mu m^3/264 \ \mu m^3 = 0.32$.

4.2 Optimum haematocrit – the viscosity of blood

4.2.1 $20/1.4 = 14.3$, $35/2.0 = 17.5$, $52/3.0 = 17.3$, $62/4.9 = 15.5$.

These ratios (centipoise/%) are graphed in Figure 2. (1 centipoise $= 10^{-3}$ N s/m^2.) The data used here are representative of those given by Pirofsky (1953). Looking just at human data, one may wonder whether the similarity between the optimum and actual haematocrits is mere coincidence, but a similar match has been found in other species too. Stone, Thompson & Schmidt-Nielsen (1968) show curves of the same kind for five mammalian species and point out that the camel has both the lowest normal haematocrit (27%) and also a 'hump' corresponding to the lowest optimum haematocrit (about 30%). There is also good correspondence between normal and optimum haematocrits in both the green turtle (Wells & Baldwin, 1994) and the rabbit (Hedrick & Duffield, 1986). In the elephant seal, however, while the optimum haematocrit is about 35%, the normal value is about 65% (higher than the 58% of question 4.1.2!), giving a high oxygen storage capacity in dives (Hedrick & Duffield, 1986). Thus, what seems to be generally an optimum may not be so in special circumstances; perhaps that applies in athletic events involving aerobic endurance.

4.3 Peripheral resistance

4.3.1   1/3, or, if allowance is made for the change in pressure, $1/3 \times 108$ mmHg/100 mmHg = 0.36.

4.3.2   As the two ventricles have the same output, only the pressures need to be compared: $(12 - 5)/100 \times 100 = 7\%$.

4.4 Blood flow and gas exchange

4.4.1   (a) $90 \text{ cm}^3/\text{s} \div 4.5 \text{ cm}^2 = 20 \text{ cm/s}$,
         (b) $90 \text{ cm}^3/\text{s} \div 4500 \text{ cm}^2 = 0.02 \text{ cm/s}$.

4.4.2 $0.05 \text{ cm} \div 0.02 \text{ cm/s} = 2.5 \text{ s}$.

4.4.3   If mixed-venous O_2 content is x, equation 4.5 gives:

$$20 \text{ l blood/min} = \frac{3000 \text{ ml } O_2/\text{min}}{(200 - x) \text{ ml } O_2/\text{l blood}}. \text{ So } x = 50 \text{ ml } O_2/\text{l blood}.$$

4.5 Arteriolar smooth muscle – the law of Laplace

4.5.1 $736 \text{ mmHg cm}^2/\text{kg-force} \times 5 \text{ }\mu\text{m}/15 \text{ }\mu\text{m} \times 3 \text{ kg-force}/\text{cm}^2 = 736 \text{ mmHg}$.

4.5.2   736 mmHg/150 mmHg = 4.9.

4.6 Extending William Harvey's argument: 'what goes in must come out'

4.6.1 $(5.00 - 4.95) \text{ l/min} \times 20 \text{ min} = 1 \text{ l}$.

4.7 The work of the heart

This topic is commonly treated graphically and in terms of single beats; the present treatment is chosen for the simplicity of the calculations.

4.7.1 $5 \text{ l/min} \times 100 \text{ mmHg} \times 0.133 = 65 \text{ J/min}$.
         $5 \text{ l/min} \times 100 \text{ mmHg} \times 0.032 = 16 \text{ cal/min}$.

4.7.2 $(18 \text{ cal/min})/(130 \text{ cal/min}) \times 100 = 14\%$, or $(73 \text{ J/min})/(534 \text{ J/min}) \times 100 = 14\%$.

4.7.3 $1000 \text{ kg/m}^3 \times (0.2 \text{ m/s})^2/2 \div 133 \text{ N}/(\text{m}^2 \text{ mmHg}) \times \text{N s}^2/(\text{kg m}) = 0.15 \text{ mmHg}$.

4.7.4 $[1 \times (0.6 \text{ m/s})^2 + 2 \times (0 \text{ m/s})^2]/3 = 0.12 \text{ m}^2/\text{s}^2$.

5 Respiration

5.1 Correcting gas volumes for temperature, pressure, humidity and respiratory exchange ratio

5.1.1 $(716 - 47) \times 0.001159 = 0.775$ ($= 0.8$ to one decimal place).

5.1.2 $(750 - 47) \times (273 + 10) / [(750 - 9) \times (273 + 37)] = 0.866$.

5.1.3 $100(1 - 0.866) = 13.4\%$.

5.1.4   Both expressions give 3.2%.

5.1.5 $(79 \text{ ml} \times 100/80)/100 \text{ ml} \times 100 = 98.75\%$.

5.1.6 $3.95/(21 - 15.8) = 0.76$.

5.2 Dissolved O_2 and CO_2 in blood plasma

5.2.1 $(0.03 \text{ mmol/l mmHg}) \times (40 \text{ mmHg}) = 1.2 \text{ mmol/l}$.

5.2.2 $(0.0014 \text{ mmol/l mmHg}) \times (100 \text{ mmHg}) = 0.14 \text{ mmol/l}$.

5.3 P_{CO_2} inside cells

5.3.1 $(56 \text{ mmHg} - 0) \times 1/20 = 2.8 \text{ mmHg}$.

5.3.2 $43 \text{ mmHg} + 2.8 \text{ mmHg} = 45.8 \text{ mmHg}$.

5.3.3 $43 \text{ mmHg} + (56 - 10) \text{ mmHg} \times 0.8/20 = 44.8 \text{ mmHg}$.

5.4 Gas tensions at sea level and at altitude

5.4.1 $[(500 - 150) \text{ ml} \times 40 \text{ mmHg} + 150 \text{ ml} \times 0 \text{ mmHg}]/500 \text{ ml} = 28 \text{ mmHg}$.

5.4.2 $333 \text{ mmHg} \times 21/100 = 70 \text{ mmHg}$.

5.4.3 $(70 \text{ mmHg} - 40 \text{ mmHg}/0.9) = 25.6 \text{ mmHg}$.

5.5 Why are alveolar and arterial P_{CO_2} close to 40 mmHg?

5.5.1   Warm water: $(160 - 0) \text{ mmHg}/25 = 6.4 \text{ mmHg}$.
Cool water: $(160 - 0) \text{ mmHg}/35 = 4.6 \text{ mmHg}$.
So the maximum is 6.4 mmHg for 0–30 °C (Rahn, 1966; Burton, 1973).

5.5.2 $0.8 \times (150 - 0) \text{ mmHg} = 120 \text{ mmHg}$.

5.5.3   (a) $2.9/2.9 = 1.0$, i.e. all of the metabolic rate!
(b) $29/2.9 = 0.1$, i.e. 10% of the metabolic rate (Burton, 1973).

5.6 Water loss in expired air

5.6.1   (a) $15,000 \text{ l} \times 47 \text{ mmHg}/760 \text{ mmHg} \times 0.8 \text{ g/l} = 742 \text{ g}$,
(b) $15,000 \text{ l} \times 37.7 \text{ mmHg}/760 \text{ mmHg} \times 0.8 \text{ g/l} = 595 \text{ g}$.

5.6.2 $742 \text{ g} - 595 \text{ g} = 147 \text{ g}$.

5.6.3 $15,000 \text{ l} \times (37.7 - 13) \text{ mmHg}/760 \text{ mmHg} \times 0.8 \text{ g/l} = 390 \text{ g}$.

5.7 Renewal of alveolar gas

5.7.1 $1 - 0.12 = 0.88$.

5.7.2 $(0.88)^2 = 0.77$.

5.7.3   Log $G = -\log 2$, so $G = 0.5$ – in accordance with the definitions of G and $t_{1/2}$.

Figure 5.3 illustrates a characteristic feature of exponential time courses (relating to the geometric progression 1.00, 0.88, 0.88^2, 0.88^3, etc.), namely a constant *proportional* change for a given *absolute* change in t. Where the exponential relationship has the form $Y = C + Y_0 e^{\pm kt}$, with C being a non-zero constant, then it is $(Y - C)$ that shows the constant proportional change for a given absolute change in t.

5.8 Variations in lung dimensions during breathing

5.8.1 $3\,l/2.5\,l = 1.2$. Subtracting dead space makes little difference.
5.8.2 $\sqrt[3]{1.2} = 1.063$.
5.8.3 $1.063^2 = 1.13$.

See Gil, Bachofen, Gehr & Weibel (1979) for changes in the microscopical appearance with lung volume.

5.8.4 $(1 - 1/1.10) \times 100 = 9\%$.

5.9 The number of alveoli in a pair of lungs

5.9.1 $(0.25\ \text{mm})^3/2 = 0.008\ \text{mm}^3$, or, more accurately, $4\pi/3 \times (0.125\ \text{mm})^3 = 0.0082\ \text{mm}^3$.
5.9.2 $(2.5 \times 10^6\ \mu l)/0.0082\ \mu l = 3.0 \times 10^8$.

From estimates of the total number of alveoli and their average dimensions, one may estimate the total alveolar surface area, and the reward is some aptly impressive figure. However, because of surface irregularities, there is much the same problem in defining the area as there is in defining the length of a coastline (Mandelbrot, 1983).

5.10 Surface tensions in the lungs

Note that Laplace's formula given here differs from that given in Section 4.5; for a cylinder, $P = T/r$, while, for a sphere, $P = 2T/r$.

5.10.1 $(15\ \text{mmHg m}\ \mu\text{m/mN}) \times (70\ \text{mN/m})/(100\ \mu\text{m}) = 10.5\ \text{mmHg}$.

The units for the '15', not given previously, are specified merely to give unit consistency in the calculation.

5.10.2 $(15\ \text{mmHg m}\ \mu\text{m/mN}) \times (25\ \text{mN/m})/(100\ \mu\text{m}) = 3.75\ \text{mmHg}$.

For a rebuttal of the entrenched idea that the surface tension in the alveoli can be zero, see Bangham (1991).

5.11 Pulmonary lymph formation and oedema

5.11.1 $25 - 7 = 18\ \text{mmHg}$. The flow would be into the capillaries.
5.11.2 $25 - 7 - 9 = 9\ \text{mmHg}$.
5.11.3 $-9\ \text{mmHg}$.

5.11.4 $9 \text{ mmHg} \times 120 \text{ }\mu\text{m}/(15 \text{ mmHg m }\mu\text{m}/\text{mN}) = 72 \text{ mN}/\text{m}$.

There is more to the roles of surfactant and of surface irregularities in the stabilizing of alveolar fluid. Guyton, Moffatt & Thomas (1984) give a good account of this interesting topic.

5.12 The pleural space

5.12.1   From capillaries to pleural space: $24 - 25 + 6 + 4 = 9$ mmHg.
5.12.2   From pleural space to capillaries: $-7 + 25 - 6 - 4 = 8$ mmHg.

Pleural liquid pressures are reviewed in Agostoni & D'Angelo (1991).

6 Renal function

6.1 The composition of the glomerular filtrate

6.1.1 $300 \text{ mosmol}/\text{kg water} \times 19.3 \text{ mmHg}/(\text{mosmol}/\text{kg water}) = 5790 \text{ mmHg}$.
6.1.2 $100 \text{ mmHg}/5790 \text{ mmHg} \times 100 = 1.7\%$.
6.1.3 $(180 \text{ l}) \times (0.01 \text{ to } 0.1 \text{ g}/\text{l}) = 1.8 \text{ to } 18 \text{ g}$.

6.2 The influence of colloid osmotic pressure on glomerular filtration rate

6.2.1 $25 \text{ mmHg}/(1 - 0.2) = 31.25 \text{ mmHg}$.
6.2.2   New net filtration pressure $= (50 - 12 - 26) = 12 \text{ mmHg}$.
        $(12 - 10) \text{ mmHg}/10 \text{ mmHg} \times 100 = 20\%$.

6.3 Glomerular filtration rate and renal plasma flow; clearances of inulin, *para*-aminohippurate and drugs

6.3.1 $(0.25 \text{ mg}/\text{min})/(4 \text{ mg}/\text{l}) \times 1000 \text{ ml}/\text{l} = 62.5 \text{ ml}/\text{min}$,
        or $(360 \text{ mg}/\text{day})/(4 \text{ mg}/\text{l}) = 90 \text{ l}/\text{day}$.
6.3.2   (a) $0.693 \times 14{,}000 \text{ ml}/(125 \text{ ml}/\text{min}) = 77.6 \text{ min}$,
        (b) $2 \times 77.6 \text{ min} = 155.2 \text{ min}$.
6.3.3 $2 \times t_{1/2} = 2 \times 77.6 \text{ min} = 155.2 \text{ min}$.

Only a small proportion of drugs exactly resemble inulin in the way they are excreted.

6.4 The concentrating of tubular fluid by reabsorption of water

6.4.1 $1/(1 - 2/3) = 3$.
6.4.2 $(125 \text{ ml}/\text{min})/(1.25 \text{ ml}/\text{min}) = 100$.
6.4.3 $100 \times 5 \text{ mmol}/\text{l} = 500 \text{ mmol}/\text{l}$.

6.5 Urea: clearance and reabsorption

6.5.1 $(450 \text{ mmol}/\text{day})/(4.5 \text{ mmol}/\text{l}) = 100 \text{ l}/\text{day}$.
6.5.2 $(1 - 70/125) = 0.44$.

6.5.3   2/3.

6.5.4 $(450 \text{ mmol})/(45 \text{ l}) = 10 \text{ mmol/l}$.

6.6 Sodium and bicarbonate – rates of filtration and reabsorption

6.6.1 $180 \text{ l/day} \times 150 \text{ mmol/l} \div 1000 \text{ mmol/mol} = 27 \text{ mol}$.

6.6.2 $27 \text{ mol} \times 58.5 \text{ g/mol} \div 1000 \text{ g/kg} = 1.58 \text{ kg}$.

6.6.3 $180 \text{ l/day} \times 25 \text{ mmol/l} = 4500 \text{ mmol/day}$.

6.6.4 $(90 \text{ mmol/day})/(4500 \text{ mmol/day}) \times 100 = 2\%$.

6.7 Is fluid reabsorption in the proximal convoluted tubule really isosmotic?

6.7.1 $$\frac{0.4 \text{ to } 4 \text{ nl/(min mm)}}{0.4 \text{ nl/[min mm (mosmol/kg water)]}}$$

$= 1\text{–}10 \text{ mosmol/kg water}$.

6.7.2   No.

For the data on rat nephrons, and for further discussion, see Schafer (1982). For more on water transport by 'osmotic engines' within cell membranes, see Zeuthen & Stein (1994).

6.8 Work performed by the kidneys in sodium reabsorption

6.8.1 $(27 \text{ mol Na/day})/(29 \text{ mol Na/mol } O_2) = \text{nearly } 1 \text{ mol } O_2/\text{day}$.

6.8.2 $1 \text{ mol/day} \times 22.4 \text{ l/mol} = 22.4 \text{ l/day}$.

6.8.3 $(22.4 \text{ l/day})/(350 \text{ l/day}) \times 100 = 6.4\%$.

6.8.4   Three times greater.

The relationship between sodium reabsorption and oxygen consumption was studied by Thaysen, Lassen & Munck (1961). There is a distinction between the total work done by the kidneys in elaborating the urine (and reflected in oxygen consumption) and the thermodynamic work needed to produce all the differences in solute concentrations between plasma and urine. The latter is only about 1% of the total. (Indeed, if the kidneys were to produce a urine identical in composition to plasma, then the thermodynamic work would be zero, but the work of reabsorption would still be substantial.)

6.8.5 $(15 \text{ ml } O_2/\text{min})/(200 \text{ ml } O_2/\text{l blood}) \times 1000 \text{ ml/l} = 75 \text{ ml blood/min}$.

6.8.6 $24 \text{ l/day} + (40 \text{ l/day})/5 = 32 \text{ l/day}$.

6.8.7 $1.25 \text{ l/min} = (16 \text{ ml/min})/(200 \text{ ml/l} - \text{venous } O_2 \text{ content})$.

Therefore, venous O_2 content $= 200 \text{ ml/l} - (16 \text{ ml/min})/(1.25 \text{ l/min}) = 187.2 \text{ ml/l}$. This is well above the oxygen content of mixed-venous blood, which is typically about 150 ml/l.

6.9 Mechanisms of renal sodium reabsorption

6.9.1 $29/5.6 = 5.2$. For glucose, the ratio is $29/6.3 = 4.6$.

6.9.2 $1/3 \times 1/6 \times 100 = 5.6\%$.

6.10 Autoregulation of glomerular filtration rate; glomerulotubular balance.

6.10.1   (a) $126 - 124 = 2$ ml/min. (b) $125 - 123 = 2$ ml/min.

6.11 Renal regulation of extracellular fluid volume and blood pressure

6.11.1 $125 - 1 = 124$ ml/min.

6.11.2 $(130 - 128)$ ml/min $\div$ $(1$ ml/min$) = 2$.

6.12 Daily output of solute in urine

6.12.1   544 to 1270 mosmol/day.

6.12.2 $(8 \text{ to } 17) \times 6.25 = 50 \text{ to } 106$ g/day.

6.13 The flow and concentration of urine

6.13.1 $(750 \text{ mosmol/day})/(300 \text{ mosmol/l}) = 2.5$ l/day.

6.13.2 $(750 \text{ mosmol/day}) \div (2.5 \text{ l/day})/2$ (or more simply $2 \times 300 \text{ mosmol/l}) = 600$ mosmol/l.

6.13.3   (a) $(750 \text{ mosmol/day})/(15 \text{ l/day}) = 50$ mosmol/l.
(b) $(750 \text{ mosmol/day})/(0.6 \text{ l/day}) = 1250$ mosmol/l.

The data of 6.13.1 – 6.13.3 lie on the curve of Figure 3 which corresponds to a constant solute load of 750 mosmol/day.

6.13.4 $2 - 0.5 = 1.5$ l/day.

6.13.5 $\dfrac{1200 \text{ mosmol/day}}{300 \text{ mosmol/l}} - \dfrac{1200 \text{ mosmol/day}}{1200 \text{ mosmol/l}} = 3$ l/day.

The previous two calculations are from Harvey (1974).

6.13.6 $(1200 \text{ mosmol/day})/(1000 \text{ mosmol/l}) = 1.2$ l/day.

6.13.7   The $(1.2 - 0.5) = 0.7$ l of extra urine in a day exceeds the 600 ml of sea water.

6.13.8 0.5 l/day $- (600 \text{ mosmol/day})/(9000 \text{ mosmol/l}) = 0.43$ l/day saved.

6.13.9   (a) $(30 \text{ mosmol/l}) \times (25 \text{ l/day}) = 750$ mosmol/day.
(b) $(50 \text{ mosmol/l}) \times (25 \text{ l/day}) = 1250$ mosmol/day.

6.14 Beer drinker's hyponatraemia

6.14.1 $(240 \text{ mosmol/day})/(6 \text{ l/day}) = 40$ mosmol/l.

For an account of beer drinker's hyponatraemia see Hilden & Svendsen (1975). Danish beer contains 1–2 mmol/l of sodium.

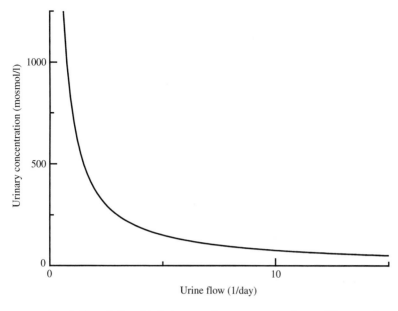

Fig. 3. The relationship between urinary concentration and flow rate for a constant solute output of 750 mosmol/day.

6.15 The medullary countercurrent mechanism in antidiuresis – applying the principle of mass balance

6.15.1 $(359.4 \text{ mosmol})/(1199.5 \text{ ml}) \times 1000 \text{ ml/l} = 299.6 \text{ mosmol/l}.$

6.15.2 $(30 \text{ ml/min} - 10 \text{ ml/min})/(120 \text{ ml/min}) \times 100 = 17\%.$ Do your textbooks suggest comparable figures?

6.15.3 $30 + 3 - 10 - 0.5 = 22.5 \text{ ml/min}.$

6.15.4 $300[(VR)_{out} - (VR)_{in}] = 300 \times 22.5 = 6750; (LH)_{out}\{300 - 100\} = 10 \times 200 = 2000; (CD)_{out}\{1200 - 300\} = 0.5 \times 900 = 450.$ The first contributes most.

6.15.5 $0 + 0.0022 \text{ ml/nosmol} \times 8300 \text{ nosmol/min} = 18.3 \text{ ml/min}.$

6.15.6 $18.3 + 22.5 = 40.8 \text{ ml/min}.$

6.15.7 $(13,800 \text{ nosmol/min})/(40.8 \text{ ml/min}) = 338 \text{ nosmol/ml} = 338 \text{ mosmol/l}.$

6.15.8 $S = [300 \times 100 + 8300] = 38,300 \text{ nosmol/min}.$
$(VR)_{out} = 100 + 22.5 = 122.5 \text{ ml/min}.$
$(38,300 \text{ nosmol/min})/(122.5 \text{ ml/min}) = 313 \text{ nosmol/ml} = 313 \text{ mosmol/l}.$

The countercurrent mechanism (reviewed by Roy, Layton & Jamison, 1992) is still incompletely understood; computer models, far more complex than the model applied here, are important in assessing rival theories.

6.16 Renal mitochondria: an exercise involving allometry

The mitochondrial content of the rat kidneys was estimated by Pfaller & Rittinger (1980). For measurements of the maximal oxygen consumption of mitochondria from mammalian skeletal muscle, see, for example, Schwerzmann, Hoppeler, Kayar & Weibel (1989).

6.16.1 $0.18 \text{ g/g} \times 2.8 \text{ g} = 0.504 \text{ g}$.

Equation 6.16 is from Edwards (1975).

6.16.2 $5.36 \times (0.35)^{0.721} = 5.36 \times 0.469 = 2.51 \text{ ml/min}$.
6.16.3 $0.32 \text{ ml } O_2/\text{min} \div 0.50 \text{ g} = 0.64 \text{ ml } O_2/\text{min per g}$.

Equation 6.17 is from Stahl (1965).

6.16.4   (a) $7.3 \times 0.35^{0.85} = 3.0 \text{ g}$. (b) $7.3 \times 70^{0.85} = 270 \text{ g}$.

7 Body fluids
A note on osmoles and osmotic pressure

The osmotic pressure of a solution depends on the total concentration of solutes in mol/kg water. In an ideal solution, 1 mol of solute in 22.4 kg of water at 0 °C exerts an osmotic pressure of 1 atmosphere, or 760 mmHg (just as 1 mol of ideal gas occupies 22.4 l when under a pressure of 1 atmosphere at 0 °C). In the case of a salt such as NaCl, the dissociated sodium and chloride contribute separately to the total.

Physiological solutions are too concentrated to show ideal behaviour and the interactions of the various solutes reduce their total osmotic effectiveness. Thus, a solution containing 150 mmol NaCl/kg water acts as if it contains only about 280 mmol of solutes (Na + Cl) per kg of water instead of 300. One way of dealing with this discrepancy is to use another unit, the osmole, such that the 'osmolality' of the solution is 280 mosmol/kg water.

Rewording an earlier statement to define the osmole, 1 osmol of solute in 22.4 kg of water at 0 °C exerts an osmotic pressure of 1 atmosphere, or 760 mmHg. The number of osmoles of a solute may be calculated from the number of moles by multiplying the latter by an empirical factor called the 'osmotic coefficient'. This varies with such things as the nature of the solute and the concentration of the solution; for NaCl in the above solution the osmotic coefficient is 280/300 = 0.93.

The osmotic pressure also increases in proportion to the absolute temperature. Thus, at 37 °C the same 1 osmol of solute in 22.4 kg of water has an osmotic pressure of $760 \times (273 + 37)/273 = 863$ mmHg. It may be useful to remember that 1 mosmol/kg water at body temperature exerts an osmotic pressure of 19.3 mmHg.

Osmoles are generally only used in the context of total solute concentration, and especially where that relates to osmotic pressure. Osmolalities are generally calculated from the colligative properties of depression of freezing point or depression of vapour pressure.

7.1 The sensitivity of hypothalamic osmoreceptors

7.1.1   (a) $[1 - 49.5/(49.5 + 0.5)] \times 100 = 1\%$.

(b) 1% of 300 = 3 mosmol/kg water.

For variations in the sensitivity of the antidiuretic hormone response to osmolality, see, for example, Robertson, Shelton and Athar (1976).

7.2 Cells as 'buffers' of extracellular potassium

7.2.1   30 mmol/15 l = 2 mmol/l. 4.5 + 2 = 6.5 mmol/l.

7.2.2   (a) 150 mmol/l × 30 l = 4500 mmol.

(b) 5 mmol/l × 15 l = 75 mmol.

7.2.3   30 mmol/4500 mmol × 100 = 0.7% – almost impossible to demonstrate.

7.3 Assessing movements of sodium between body compartments – a practical difficulty

7.3.1   (150 + 3) mmol ÷ (1 + 0.027) kg water = 149 mmol/kg water.

7.4 The role of bone mineral in the regulation of extracellular calcium and phosphate

For the four forms of calcium phosphate, see Driessens, van Dijk & Verbeeck (1986). For the solubility relations of octocalcium phosphate, see Driessens, Verbeeck & van Dijk (1989).

7.4.1   9/4.5 = 2, 9/7 = 1.29, 8.5/4.5 = 1.89, 8/6 = 1.33. The range is thus 1.29 to 2.0.

7.4.2   The concentration of free calcium would rise by $1.5 \times (1.3 - 1.0) =$ 0.45 mmol/l. If it started at 1.3 mmol/l, the percentage rise would be $(0.45/1.3 \times 100) = 35\%$, much as for phosphate.

7.4.3   The concentration of calcium would rise by $(1.5 \times 0.1$ mmol/l × 30/100) = 0.045 mmol/l, i.e. 0.45% of 10 mmol/l.

7.4.4   10 mmol/l – 1.5 × 0.1 mmol/l = 9.85 mmol/l.

7.5 The amounts of calcium and bone in the body

For the calcium content of a 70-kg body a round-number figure of 1 kg is used. Cohn, Vaswani, Zanzi & Aloia (1976) give averages for men and women between the ages of 30 and 39 that correspond, respectively, to 996 and 988 g of calcium per 70 kg.

7.5.1   1 kg × 100/26 = 3.8 kg

7.5.2   3.8 kg/70 kg × 100 = 5.4%

Since this answer is for dry fat-free bone, the proportion of bone in a living body is somewhat larger, and the proportion of skeleton, with water, fat, and other components of marrow, is even greater.

7.5.3   1.4% × 100%/40% = 3.5%.

7.6 The principle of electroneutrality

7.6.1 $\dfrac{10^{-6}\,\text{farad}/\text{cm}^2 \times 0.07\,\text{V}}{96490\,\text{coulombs}/\text{mol}} \times \dfrac{\text{coulomb}}{\text{farad} \times \text{V}} = 7 \times 10^{-13}\,\text{mol}/\text{cm}^2$
(0.7 picomole/cm^2).

7.6.2   (a) $0.028/7 = 4 \times 10^{-3}$ mmol/l.
(b) $0.028/1 = 28 \times 10^{-3}$ mmol/l.

7.6.3 $(144 + 4 + 2 \times 1 + 2 \times 0.5) - (102 + 28 + 1 + 18) = 2$ mequiv/l.

In relation to the accurate analysis of plasma, remember that the concentrations of bicarbonate and chloride change with carbon dioxide tension, along with the net charge of the plasma proteins, through buffering and through chloride/bicarbonate exchange across erythrocyte membranes. It is sometimes forgotten that all three differ as between arterial and venous plasma, and in blood samples exposed to air. The total concentration of chloride plus bicarbonate stays nearly constant, except inasmuch as protein ionization changes and small shifts of water occur between erythrocytes and plasma.

7.6.4 $(138 + 3 + 3 + 6) - (70 + 5) = 75$ mequiv/l.

7.6.5 $18 + 135 + 0.5 \times 2 - 78 - 16 = 60$ mequiv/kg water.

There is clearly a large quantity of anions not accounted for. These include protein (mainly haemoglobin) and phosphates (e.g. 2,3-diphosphoglycerate). The calculation cannot reveal whether or not other cations are present.

7.7 Donnan equilibrium

7.7.1 $141.3 + 18 = 159.3$ mmol/kg water.

7.7.2 $150/159.3 = 141.3/150 = 0.942$.

7.7.3 $61.5\,\text{mV} \times -0.026 = -1.6$ mV.

7.8 Colloid osmotic pressure

7.8.1 $300\,\text{mosmol}/\text{kg} \times 19.3\,\text{mmHg}/(\text{mosmol}/\text{kg}) = 5790$ mmHg.

7.8.2   Solution 2, by 0.6 mmol/kg water.

7.8.3 $1 + 0.6 = 1.6$ mosmol/kg water.

7.8.4. $1.6\,\text{mosmol}/\text{kg} \times 19.3\,\text{mmHg}/(\text{mosmol}/\text{kg}) = 31$ mmHg.

7.8.5 $1\,\text{mmol}/\text{kg water} \div 1000\,\text{mmol}/\text{mol} \times 68{,}000\,\text{g}/\text{mol} = 68$ g/kg water.

For more on colloid osmotic pressures and protein net charge, see Burton (1988).

7.9 Molar and molal concentrations

7.9.1   Equation 7.10: $0.99(1 - 0.75 \times 70/1000) = 0.94$ kg water/l.

7.9.2   With w from calculation 7.9.1, $(141\,\text{mmol}/\text{l})/(0.94\,\text{kg water}/\text{l}) = 150$ mmol/kg water. 150 is notably different from 141.

7.9.3 $0.99(1 - 0.75 \times 360/1000) = 0.72$ kg water/l.

7.9.4 $150\,\text{mmol}/\text{kg water} \times 0.72\,\text{kg water}/\text{l} = 108$ mmol/l – a much greater discrepancy than for plasma.

7.10 Osmolarity and osmolality

7.10.1   281.5 mmol/l.

7.10.2 $0.93/0.94 = 1.0$ to one decimal place.

7.11 Gradients of sodium across cell membranes

7.11.1   10 to 13 kcal/equiv ÷ 23.1 kcal/volt equiv ÷ 3 × 1000 mV/V = 144 to 188 mV.

7.11.2   Somewhat above $(90 + 40) = 130$ mV.

7.11.3 $61.5 - (-90) = 151.5$ mV.

7.11.4   Log 2 = 0.30, as stressed in Appendix B. $61.5(1 + 0.30) - (-90) = 170$ mV.

7.11.5 $61.5 - \dfrac{2}{3} \times (-94) - \dfrac{1}{3} \times (-90) = 154$ mV = 0.154 V.

7.12 Membrane potentials – simplifying the Goldman equation

7.12.1   (a) $0.05 \times 15/150 \times 100 = 0.5\%$

          (b) 5%, likewise.

7.12.2   (a) $61.5 \log(4.5/150) = -93.7$ mV.

          (b) $61.5 \log [(4.5 + 1.5)/150] = -86.0$ mV.

          (c) $61.5 \log [(4.5 + 0.07 \times 150)/150] = -61.5$ mV.

8 Acid–base balance

8.1 pH and hydrogen ion activity

8.1.1 4×10^{-8} mol/l or 40 nmol/l. This can be obtained using log 2 as follows: $10^{-7.4} = 10^{-8} \times 10^{0.6}$; $0.6 = 2 \log 2 = \log 4$, so that $10^{0.6} = 4$; $10^{-8} \times 10^{0.6} = 10^{-8} \times 4$.

8.1.2   1 μm^3 contains $10^{-7} \times 6.0 \times 10^{23} \times 10^{-15} = 60$ hydrogen ions.

8.1.3 $3 \times (0.1 \ \mu m)^2 \times 4 \ \mu m = 0.12 \ \mu m^3$.

8.1.4 $60 \times 0.12 = 7$.

8.2 The CO_2–HCO_3 equilibrium: the Henderson–Hasselbalch equation

For more on the equation, and on the meaning of pK_1' see the Notes for Section 1.5. pK_1' is taken as 6.1 throughout this book, but, for accurate work, various complications need to be considered (Burton, 1987).

8.2.1   (a) pH $= 6.1 + \log (12/1.2) = 7.1$.

          (b) pH $= 6.1 + \log (24/1.2) = 6.1 + \log 10 + \log 2 = 7.4$.

8.2.2   It falls by 0.3 unit.

8.2.3   It falls by about 0.3 unit, the point being that 83/39 is close to 2, so that $\log (83/39) = $ ca. 0.3 (actually 0.33).

8.2.4 $6.1 - \log 0.03 = 7.62$.

8.2.5 $7.4 - \log (24/40) = 7.62$. This answer has to be the same as the previous one, because the three variables have normal and compatible values.

8.2.6 $(1.023 - 1) \times 100 = 2.3\%$.

8.2.7 $0.05 \times 0.2 = 0.01$. Therefore the answer is again 2.3%.

There are other ways of graphing acid–base data, notably $[HCO_3]$ against pH. The graph of $[HCO_3]$ against P_{CO_2} has the advantages of utilizing as axes the two determinants of pH and yielding for plasma *in vitro* a curve that relates to the carbon dioxide dissociation curve.

8.3 Intracellular pH and bicarbonate

8.3.1 $61.5\ \text{mV} \times (7.0 - 7.4) = -24.6\ \text{mV}$.

8.3.2 $-24.6 - (-70) = +45.4\ \text{mV}$.

8.3.3 $61.5\ \text{mV} \times \{pH_i - pH_e\} = -70\ \text{mV}$. $\{pH_i - pH_e\} = -1.14$. $pH_i = 7.4 - 1.14 = 6.26$.

8.3.4 $26\ \text{mmol/kg water} \div 2 = 13\ \text{mmol/kg water}$.

8.3.5   1.0; the two quantities are equal.

8.4 Mitochondrial pH

8.4. 1 $6.1 + \log \{150/(0.03 \times 45)\} = 8.15$.

8.4.2   Halving $[HCO_3]$ lowers the pH by log 2 ($= 0.30$) to 7.85.

8.5 Why bicarbonate concentration does *not* vary with P_{CO_2} in simple solutions lacking non-bicarbonate buffers

8.5.1 $(24.00000 + 0.00010 - 0.00004) - (0.00005 - 0.00008) = 24.00009$ mmol/l. $[HCO_3]$ is raised, but not by a measurable amount.

8.5.2 $[HCO_3]$ is virtually unchanged, but $[H]$ is doubled. Therefore, by application of the Henderson–Hasselbalch equation, $P_{CO_2} = 2 \times 40\ \text{mmHg} = 80\ \text{mmHg}$.

8.6 Carbonate ions in body fluids

8.6.1   (a) $10^{(7.8-9.8)} = 0.01$; (b) 0.1; (c) 1.0.

8.6.2 $7.62 + \log (20/1.3) = 7.62 + \log 15.4 = 8.81$.

8.7 Buffering of lactic acid

8.7.1   (a) $10^{(4.6-6.6)} = 0.01$. b) $10^{(4.6-7.6)} = 0.001$.

8.7.2 $7.4 - \log (25/20) = 7.3$.
         Note that $\log (25/20) = [\log 10 - \log 2] - 2 \log 2 = 0.1$.

8.7.3 $pH = 6.1 + \log (20/6.2) = 6.61$.
         Alternatively, the previous answer minus $\log 5 = (7.3 - 0.7) = 6.6$.

8.8 The role of intracellular buffers in the regulation of extracellular pH

8.8.1   The pH rises by log $(28/25) = 0.05$.

8.8.2 $7 \times 3/15 = 1.4$ mmol/l.

That the amount of bicarbonate leaving the erythrocytes is influenced by the final concentration in the extracellular fluid, and therefore by other sources of bicarbonate, may be understood by analogy. If a warm object ($\equiv$ erythrocytes) is dropped into water ($\equiv$ extracellular fluid) that is cooler, the extent to which the object loses heat ($\equiv$ bicarbonate) is greater if the volume of water is greater. The object loses less heat if other warm objects ($\equiv$ nucleated cells that also release bicarbonate) are dropped in with it.

8.8.3   A factor of 2 also.

8.8.4   None.

For a more detailed discussion of the movements of bicarbonate between cells and extracellular fluid in disturbances of acid–base balance, see Burton (1992).

8.9 The role of bone mineral in acid–base balance

8.9.1 $25 + 1 = 26$ mmol/l.

8.9.2   A rise in pH of log $(26/25) = 0.017$ unit.

Buffering by bone mineral has been reviewed by Green & Kleeman (1991) and by Burton (1992).

8.10 Is there a postprandial alkaline tide?

The figures for the rates of gastric acid secretion are from Bowman & Rand (1980); pharmacology books are the readiest sources of such data, because of the relevance of the data to antacids. Does the size of a typical antacid tablet (sodium bicarbonate, perhaps) seem right for those rates? Johnson, Mole & Pestridge (1995) were unable to demonstrate either a renal or a respiratory alkaline tide in response to gastric acid secretion.

8.10.1 25 mmol/l $+ (28$ mmol$)/(14$ l$) = 27$ mmol/l.

8.10.2   Log $(27/25) = \log 1.08 = 0.033$ pH unit.

9 Nerve and muscle

For more on 'number numbness', see Hofstadter (1985) who taught that physics class.

9.1 Myelinated axons – saltatory conduction

9.1.1 $(1.5$ mm$)/(100$ mm/ms$) = 0.015$ ms.

Some of those textbook figures depict action potentials in squid axons or mammalian nerve fibres at low temperature, which are therefore slower, but this does not explain the difference.

9.1.2   At least 1.5 m.

9.1.3 $(1 \text{ m})/(10 \text{ m/s}) \times 1000 \text{ ms/s} - 10 \text{ ms} = 90 \text{ ms}$.

9.2 Non-myelinated fibres

9.2.1 $120^2/4.8 = 3000 \ \mu\text{m (3 mm)}$.

It is not known whether the myelinated fibres in the central nervous system with diameters of 0.2–1 μm conduct more or less rapidly than non-myelinated fibres of the same diameter.

9.2.2 $2.2 \times \sqrt{100} = 22 \text{ m/s}$.

9.3 Musical interlude – a feel for time

9.3.1 $1000 \text{ ms/s} \div 11/\text{s} = 91 \text{ ms}$.

9.3.2   The reaction to a sound would probably take between $(150 - 50) = 100$ ms and $(250 - 30) = 220$ ms, both times being longer than the previous answer.

9.4 Muscular work – chinning the bar, saltatory bushbabies

The method of calculating the volume of the muscle by treating it as a set of thin slices is essentially that proposed by the astronomer Johannes Kepler in 1615 for estimating the volume of wine in a barrel. Physiologists use essentially the same method to estimate the volumes of microscopical structures that are present in successive serial sections of tissue.

9.4.1 $9.8 \text{ J/kg m} \times 0.4 \text{ m} = 3.9 \text{ J/kg body mass}$.

9.4.2 $3.9/70 \times 100 = 5.6\%$.

The muscles, kindly dissected and weighed by Dr. A. Chappell, were the latissimus dorsi, teres major, pectoralis major (less the clavicular head), biceps, coracobrachialis, brachialis and brachioradialis.

9.4.3 $5.6\%/2 = 2.8\%$

Another route to the conclusion that small people should generally find it easier to lift themselves than do large ones is as follows. The ease of holding the body up after a lift increases with the tensions sustainable by the relevant muscles, and therefore their cross-sectional areas, and it decreases with body weight. For a body of given build (i.e. muscularity and relative proportions), areas (including those of muscle cross-section) are proportional to the square of any given linear dimension 'L', i.e. to L^2, while body weight is proportional to body mass, hence to volume and to L^3. The ease with which one supports oneself is therefore proportional to $L^2/L^3 = L^{-1}$. As to the lift, two other factors are relevant, the height of the lift and, opposing this, the lengths of the relevant muscles. Both are proportional to L. Ease of lifting is therefore proportional to $L^{-1} \times L/L = L^{-1}$.

9.4.4 $2.1 \text{ m} \times 9.8 \text{ J/kg m} = 20.6 \text{ J/kg}$.

9.4.5 $20.6 \text{ J/kg body} \div 70 \text{ J/kg muscle} \times 100 = 29.4\%$.

The calculations on the bushbaby derive from a paper on muscle work by Alexander (1992) and references therein.

9.5 Creatine phosphate in muscular contraction

9.5.1   70 J/kg ÷ 50 J/mmol = 1.4 mmol/kg.

9.6 Calcium ions and protein filaments in skeletal muscle

9.6.1   (a) $\pi \times (0.5\ \mu m)^2 = 0.79\ \mu m^2$.
         (b) $1\ \mu m \times 0.79\ \mu m^2 = 0.79\ \mu m^3$.

9.6.2 $0.8\ \mu m^3 \times 10^{-15}\,l/\mu m^3 \times 10^{-7}\ mol/l \times 6 \times 10^{23}$ ions/mol = 48 ions.

9.6.3 $0.8\ \mu m^2/(8.7 \times 10^{-4}\ \mu m^2) = 920$.

9.6.4 $920/48 = 19$.

9.6.5 $47{,}520 \times 4/10{,}000 = 19$.

9.6.6 $3 \times (450\ nm)/(45\ nm) = 30$.

Appendix B: Exponents and logarithms

1    $10 \log 2 = $ nearly $3 \log 10 = $ nearly 3.00. $3.00/10 = 0.300$ (too low by 0.001).

2    $3 \times 0.301 = 0.903$.

3    $1 - 0.301 = 0.699$.

4    $(2 - 0.301)/2 = 0.85$. Log 7 is actually 0.845.

INDEX